Falling
for
Snow

A Naturalist's Journey into the World of Winter

Jamie
Bastedo

Red Deer
PRESS

The Publishers
Red Deer Press
813 MacKimmie Library Tower
2500 University Drive N.W.
Calgary Alberta Canada T2N 1N4

Credits
Edited for the Press by Jill Fallis
Cover design by Duncan Campbell
Text design by Erin Woodward
Cover photograph courtesy of Ryan McVay / Getty Images / Photodisc
Printed and bound in Canada by AGMV Marquis for Red Deer Press

Acknowledgements
Financial support provided by the Canada Council, the Department of Canadian
Heritage, the Alberta Foundation for the Arts, a beneficiary of the Lottery Fund of the
Government of Alberta, and the University of Calgary.

National Library of Canada Cataloguing in Publication Data
Bastedo, Jamie, 1955–
Falling for snow : a naturalist's journey into the world of winter /
Jamie Bastedo.
ISBN 0-88995-265-5
1. Snow. 2. Snow ecology. I. Title.
QH541.5.S57B37 2003   577.5'86   C2003-910059-6

5 4 3 2 1

To Sri Sathya Sai Baba
For the magic

and mystery
of snow.

Contents

Acknowledgements

This project germinated many years ago when a local cub scout leader asked me if I could "do something on snow" for his pack of kids. The request reminded me that I had fallen into the comfortable path of a "seasonal naturalist," getting out in the field mostly in spring and summer, then spending the rest of the year indoors writing about birds, flowers, rocks, forest fires—things that have nothing at all to do with winter. Till then, I had given short shrift to an elemental part of nature which dominates my neck of the north woods for over half the year: snow. With little more to draw upon than the snowdrifts out my window, I shrugged and told the guy I'd do it.

A few days later I was out behind the local museum with a bunch of giggling ten-year-olds, practicing Inuktitut snow words, doing snow dissections with a chilled spatula and frying pan, and climbing all over a barn-sized *quinzhee* snowhouse (well, it *seemed* pretty big to the kids). So, to that scout leader, whoever he was, who called me up for my first snow gig, I say: thanks. You helped me reopen a door into the playful world of snow.

Much of the initial snow research that fed this book was sponsored by video producer Ross Burnet for a two-part series called "Ernie's Earth: The Wonders of Winter." Thanks for that kick-start, Ross. From the outset, I got the strong impression that accessible *and* readable publications on snow were scanty at best. For helping me dig up deeply buried treasures in this apparently obscure literature, I thank Don Gray, Brian Latham, Jack Ledger, John Pomeroy, Bill Pruitt, Bob Reid, and all the library staff at the Arctic Institute of North America in Calgary.

Of the authors I discovered among these treasures, I am particularly grateful to Duncan Blanchard, Andy Goldsworthy, Monte Hummel, Richard

Kucera, Edward LaChapelle, Blake McKelvey, Bernard Mergen, Bill Pruitt, Doug Sadler and Chet Raymo. I trust I have given your wonderful words on snow their due honor. I am also very grateful to all the authors, poets and publishers who permitted me to grace the following pages with excerpts from their material. An exhaustive effort was made to locate all rights holders and to clear reprint permissions for all major quotations. If any required acknowledgements have been omitted, or any rights overlooked, it is unintentional and forgiveness is requested. If notified, the publisher will be pleased to rectify any omission in future editions.

Thank you to all the photographers who graced my words with their pictures, particularly Fran Hurcomb, Tessa Macintosh, John Poirier, Lorne Schollar, and the editor at Northern News Services.

Among the many who shared their snow experiences with me, I thank Arthur Boutilier, Miles Davis, John Lee, Damian Panayi, Chris Straw, Ted Warren, my mother, and the Parks Canada and Brewster staff at the Athabasca Glacier. And, oh yes, the anonymous iceman whose glacier tour I intend to take next time—as long as he goes first.

Thank you, Snowman Bill Pruitt, for giving the manuscript a thorough technical drubbing and for your sage advice on "how to write goodly." Thanks everyone at Red Deer Press for shepherding the manuscript to completion, particularly managing editor Dennis Johnson for your lucid editorial guidance, Jill Fallis for your impeccable copyediting, and production managers Daniel Krut and Erin Woodward for the whip cracking that helped deliver this baby to the maternity ward.

Finally, thank you to Brenda, Jaya, and Nimisha for once again lovingly supporting my writer's habit, especially on those days when an abominable blizzard seemed to escape from my office every time I opened the door.

It is extremely improbable that anyone has as yet found, or, indeed, ever will find, the one preeminently beautiful and symmetrical snow crystal that nature has probably fashioned when in her most artistic mood.

—Snowflake Man Wilson A. Bentley[1]

Foreword

Dear Reader,

There's a hoary old legend that says the Inuit have 24 (or is it 36? or 48?) different words for snow. I'm not impressed. I personally have a couple of dozen words for the stuff—none of them printable in a family-friendly book like this.

Speaking of which—a Northern Canadian writing a book about the wonders of snow? Isn't that kind of like a Saudi Arabian writing about the wonders of sand? Well, it could be—but not in the hands of Jamie Bastedo. He is that rarity among North Americans: a genuinely sane citizen who really, truly loves snow.

He's fascinated by the stuff. You and I, we curse and grumble as we shovel out our driveway only to watch the snowplow fill it in again. Jamie told me he has actually been late for work because he paused to stare at the snow crystals piling up on his windshield. I believe him.

A few years back, Jamie cornered me in downtown Yellowknife, Northwest Territories, and asked me if I wanted to see something wonderful. Sure, I said. We hopped in his truck, drove out to the deep boonies, hiked down a trail, and came to a clearing. Jamie went over behind a pine tree and came back with a frying pan and a spatula. I explained that I'd already had breakfast. Jamie said that we were not eating. We were going to do a snow dissection.

If anyone had ever told me I would one day find myself standing in the wilderness in midwinter, listening to a man expound on the infinite variety and boundless beauty of the white stuff, I'd have dismissed the oracle as a raving nutbar, but there I was.

And you know what? I was entranced by Jamie's "Snowflakes 101" lecture. When it comes to his favorite topic, Jamie has the evangelical power to convert the most jaded and winter-weary snow foe into a true believer.

So reader, think twice before you turn to page one. You could be embarking on a whole new love affair.

Now, don't forget your mittens.

Arthur Black
Author, Humorist, Broadcaster

Introduction
Notes From a Snowhouse

The winter is thrown to us like a bone to a famishing dog,
and we are expected to get the marrow out of it.

—Henry David Thoreau[2]

A s a rule, I generally avoid buying the local newspaper, or any newspaper for that matter. I have my reasons. First, most papers I do buy go virtually unread into my woodstove. How some people find the time, or reason, to digest those bulky tabloids on a daily basis is beyond me. And second, there seems to be little news that's fit to print these days, good news anyway. But occasionally, in a weak moment, I break down and toss our local rag into my shopping cart, sucked in by some especially provocative image or headline plastered over the front page.

A few years ago, what steered my unthinking hand to the newspaper rack was an image of my home town of Yellowknife, Northwest Territories, nestled under a massive skintight dome to shelter it from the long subarctic winter. Standing larger than life behind the dome was a grinning man with arms outstretched, sporting a fedora and dark glasses. In my store-bound somnolence, I presumed this was some southern developer who had come north to peddle his wacky ideas. (He certainly would not be the first.) It was back home, after a strong cup of tea, that I connected this bizarre image with the date: April 1. Sucked in again, Jamie.

The newspaper's editors got a good laugh over this Fool's Day farce, especially when lots of readers started phoning in to the newsroom to ask if

Unfettered by sky domes, winter takes charge of my subarctic Yellowknife home.
Tessa Macintosh

that developer dude was for real, or how soon would the dome be built, or could it maybe be stretched over their neighborhood? The fact that a hick little newspaper could pull the wool over so many people's eyes, including mine, suggests that the idea of sealing ourselves off from all that winter stands for—numbing cold, icy winds, and snow—lies not far beyond the realm of the credible. Across North America's snowbelt, legitimate proposals for domed communities have dropped on and off city planners' drawing boards for more than forty years. Though the idea smacks of science fiction, there are folks out there who would gladly throw their faith at it, aching to say good riddance to winter.

Credible, maybe. Desirable? Not in my books. Call me flaky, but I happen to love winter. While living in India for over a year, what I ached for most was not flush toilets or phones that actually worked or "Hockey Night in Canada," but snow. And not some quaint dusting of snow to playfully nip my nose, then vanish an hour later, but a wild and woolly snowscape that's unequivocally and predictably in my face for the better part of the year. Wrap my town in a snow-shedding skin of Plexiglas and I'm outta here. The only dome I want over my head is one made of snow itself.

A couple weeks after that devious newspaper story hit the stands, I sat alone in a *quinzhee,* my favorite brand of snowhouse, scratching away in my snow journal with a blobbing ballpoint pen less comfortable with the cold than I.

> *My pen is dying in the cold. But it really isn't so bad in here. That's the whole idea. The frigid snow all around keeps me warm. Though the sun is hot and high this Easter Sunday, my backpack thermometer tells me it is eight degrees warmer inside than out. But another few days of this sun and my frosted cocoon will have to succumb to spring, caving in with a dull thump when nobody's looking.*
>
> *In the meantime, here I sit, on a soft, insulating caribou skin, listening to the muffled laughter of my girls tobogganing over a small cliff in front of our cabin and to the drips dancing in syncopated time off plywood planks by my sunny door. Before it all melts, I look at snow, I think of snow, I say good-bye to snow.*
>
> *Snow above, snow below, snow all around me. I can still see remnants of the original pukak crystals, scooped two months ago from deep virgin snow to build this shelter. Time and the steam of our overnight breaths have weathered most of the crystals' delicate lattice shapes into amorphous blobs. I tap them with my knuckles, and my eyelids fill with frost. My view of individual snow grains is best when I face south, where the slowly thinning dome now admits a warm turquoise glow.*
>
> *The vent hole above me, once choked with fresh dry snow, is now a fast widening skylight through which I can view the naked tops of poplars and white spruce against an azure sky.*
>
> *Haley [my dog] has just come in for a visit, licking my face, lying down on my sleeping bag for a quick scratch behind the ears, then woofing at me to come out and play in the snow. At first she was timid to enter the quinzhee, as were some of the smaller humans in camp. But now she knows that she is always welcome in my warm house of snow.*

I live in a far northern land, where what most people would call winter lasts for more than six months of the year. Here in Canada's Northwest Territories on the rocky north shore of Great Slave Lake, the odds of a brown Christmas are roughly at par with the odds of a white Christmas in Fiji. As an ardent *chionophile* (snow lover), I admit there are some practical tradeoffs to living in the subarctic—like having to design my kids' Halloween

At home in the snow: winter camper surfaces from his cozy quinzhee snowhouse.
Tessa Macintosh

costumes to fit over a snowsuit or packing a parka for a late May canoe trip.
But I feel lucky to live in a land of snowy superlatives. Our snow is so dry I
can wander through it for hours without getting soaked. It's so fluffy I can
dive into a snowbank as easily as falling into a hammock. It's so cold I can
build a rock-hard quinzhee in no time. And wherever the snowcover lasts a
long time—as in any northern or alpine environment—it gets a chance to
fully ripen over the winter, creating a fascinating kaleidoscope of layers,
shapes, and textures that I can crack open anytime, anywhere, and read like
a book.

But to unlock the secrets of snow, you don't have to live in a log cabin
north of sixty or an alpine hut near the treeline. Whether your snow lasts six
days or six months of the year, it deserves a close second look, because out
on the trail a grounding in snow lore can boost the fun, comfort, traveling
ease, and safety of your next winter trek—even if it's in an armchair with a
book like this.

A lot of people have been telling me lately, "You'd better get that
thing finished before it gets much warmer." What they are talking about is
not an ice-cream sandwich, which can get a little messy in the heat, but this
very book. Honestly, the thought has occurred to me that *Falling for Snow's*
potential readership might steadily decline as global temperatures creep
upwards due to our heedless meddling with the climate. I hear more and

more stories from across Canada and the snowy states of America about fleeting winters with scanty snow. I listen to friends down south complain that their skis and snowboards sit idle in the garage for most of the winter. I learn that even the Inuit snow lords are having a tough time building igloos these days due to fickle and fast-changing snow conditions. And I read with genuine alarm that snowcover in the northern hemisphere has been shrinking over the past 25 years (some people might react differently to this news). But then another whopper snowstorm blankets Toronto or Chicago, and I am heartened that snow retains its perpetual relevance, even for those who live on the outer limits of North America's snowbelt. That being so, I have written the following seventy thousand words in celebration of the many faces of snow.

What is this white stuff, anyway? It charms the soul yet threatens the body; it displays exquisite fragility yet can carve continents; it eternally shrouds our highest mountains yet embodies all in nature that is transitory. In Chapter 1, "Stalking Snow," I tell the story of my pilgrim's journey to North America's most famous glacier to probe some of the mysteries and paradoxes of snow. Show me any snow crystal. Its life cycle abounds with riddles along every step of the way—how it forms, how it falls, how it accumulates, how it changes on the ground, and ultimately how it disappears. Chapter 2, "Being Snow," explores these riddles through the writings and musings of snow scientists and an intimate biography of one crystal's unique story.

Chapter 3, "Living Snow," explodes the myth that snow is a changeless, deathlike shroud over the landscape. Through private portraits of the lives of snow dwellers—big and small; under, in, and on the snow—I show how snow is, in fact, a dynamic life-supporting blanket that has left a profound imprint on the natural history of North America. My reckoning is that no other element in nature provokes such wildly contrasting emotions as snow. By drawing on my favorite snow poems and stories, Chapter 4, "Thinking Snow," takes a free-ranging look at this colorful palette of perceptions through different cultural and historical lenses.

Nothing tests a town's mettle more than a thumping good snowstorm. This drama has been played out over and over again, throughout history and on many fronts. Chapter 5, "Confronting Snow," tells the story of how cities across North America have, or have *not*, adapted to snow's biggest punches. How we turn to snow as a primary source of freedom and fun is the focus of Chapter 6, "Playing in Snow." From snowshoeing to kite-skiing, those of us lucky enough to live in the snow belt are prodigious players in snow— and always will be. Finally, in the epilogue, "Seeing Snow," I share a personal journey with the Inuit, snow lords of the arctic, who, to me, prove just how far we humans can go in bonding our lives and hearts to snow.

Snow—you've got to love it!
Tessa Macintosh

So there you have it. There is no instruction manual provided with this book. Where and how you read it is your business. You may choose to randomly flip through it in a candle-lit quinzhee built by your own hands (if you're serious about this, see the Appendix for tips on how to build one). Or you might read it from cover to cover while parked by a crackling fire in a penthouse thirty storeys up. However you relish the pages that follow, my hope is that the next time the white stuff falls, you will never again be tempted to greet it as "plain old snow." I promise you that once you get over that hump, you can't help but fall for it.

Stalking Snow

Nature was here something savage and awful, though beautiful. I looked with awe at the ground I trod on, to see what the Powers had made there, the form and fashion and material of their work . . . There was clearly felt the presence of a force not bound to be kind to man.

—Henry David Thoreau, "Ktaadn"[3]

Glimpsing Goliath

You have to hide something almost a hundred times bigger than Manhattan's Central Park. Where do you put it? Deep-six it in the ocean? Fire it, piece by piece, to the back side of the moon? Or, how about sticking it inside a lofty mountain basin rimmed by eleven of the highest peaks in the Canadian Rockies? That's where a couple of mountaineers, who weren't even looking, found Alberta's Columbia Icefield.

This alpine ocean of snow and ice is, in places, more than 25 kilometers (15 miles) wide and 300 meters (1,000 feet) thick. Hard to miss, you would think. But, tucked away as it is, in utter desolation, it is hailed as the last major geographical feature discovered in the Canadian west. Ironically, over the past fifty years it has become one of the most visited natural wonders in North America. On a good day—that is, when there are no blizzards, whiteouts, or smothering fog—this elusive feature, measured by raw tourist numbers, probably beats out the Grand Canyon and Niagara Falls combined. The Columbia Icefield is, I discovered, a place of paradoxes, not the least of which is the very snow that keeps this monster alive.

In truth, the real drawing card isn't the icefield itself but the ring of glaciers that spill down from it like arms on a colossal octopus. Reaching lengths of more than 6.5 kilometers (4 miles), these tongues of ice drool water to three oceans, dousing much of western Canada along the way. Grandest of them all is the Athabasca Glacier, the most gawked-at glacier in the world. But, though I call myself a naturalist, visit Alberta often, and lovingly decorate my yard with glacier-ground rocks, I had never set foot on this most famous of glaciers. Until now, that is.

❄ ❄ ❄

I know my tourist days are numbered when I start reflexively looking at my watch. For the past two weeks I haven't even worn one. But now our family road trip through Alberta is winding down, and we can spare only part of a morning to "do the glacier" before pushing north for home.

First stop: the Columbia Icefield Centre, a grand tourist Mecca built on a rubble pile donated by the Athabasca Glacier. Outside, this hilltop hacienda presents a classic Parks Canada face, complete with green gables, dormer windows, and a meticulous limestone jacket. Inside, it feels and looks like a cross between a shopping mall and an airport. On this mid-August morning, the place is buzzing.

For wallpaper, the Icefield Centre uses flashy posters that advertise the reason mobs of people come here: to go for a joyride on the Athabasca Glacier. And not in their cars or in a school bus or—heaven forbid—on foot, but on an original, bona fide, nowhere-else-on-earth-can-you-do-this Brewster Snocoach. Each one costing more than half a million bucks and carrying almost sixty passengers, these monstrous six-wheel-drive behemoths were engineered exclusively for glacier gawkers like me.

"Discover the ice age," says one poster. "Take an unforgettable odyssey through time," says another. Or my favorite: "You've come this far. Do the ice!" More to the point: "Wow!" And in smaller print: "Departures every 15–30 minutes."

All the posters promise unique thrills—but not cheap ones. A young woman at the Brewster check-in counter says it will cost me over eighty bucks to get my family of four onboard the Snocoach. "Worth every penny," she adds, while swiping my well-worn credit card.

"So," I say, flipping out my dog-eared note pad. "Just how many people *do* the ice on a typical day?"

She punches my query into her computer. "Since April we've toured . . . let's see . . . about 580,000 people."

"Excuse me?"

She speaks up, as if through a language barrier. "Five hundred and eighty thousand. Our target for October is 638,000."

I glance at the swelling crowd, then back at her. "And how many is that a day?"

"It goes up and down. Maybe three to five thousand." She hands me my tickets. "Please report to Gate B in 30 minutes. Have a nice day."

What I hear is, "Have an *ice* day."

So we and a small town's worth of people are about to embark on "the unique experience of a lifetime." My two girls, aged nine and eleven, are hungry. I send them off with my wife, Brenda, to rustle up some midmorning French fries—I told you already, we're on vacation. It's time for me to hit the gift shop.

I have a chronic allergy to gift shops. They make me break out in shivers of disdain and mild convulsions of contempt. This affliction is often tempered by my obsessive-compulsive appetite for books. Almost on tiptoe, I enter the Icefield Centre's sprawling gift shop, being careful not to knock over any fake soapstone polar bear carvings or tawdry totem poles with my lumpy backpack. Between me and the bookshelves is a display of undersized beer mugs and oversized pencils, both adorned with archetypical snowflakes and both equally useless. For some reason, I hold my breath as I walk past them.

Ahh, the books. Their glossy covers are adorned with picture-perfect mountains, glaciers, alpine flowers, and cuddly grizzly cubs. I discover that most of the books are in Japanese or German. After some digging I find the English section, where several glacier guidebooks jump off the shelf at me. As I scoop them under my arm, I am already salivating.

I park myself against a trinket-less wall and start thumbing through *Exploring the Columbia Icefield,* by glacial guru and geologist Richard Kucera. For kicks, this man goes hiking on the Greenland Ice Sheet or photographing glaciers that hang from the slopes of Mexico's highest volcanoes. One of his pet projects is making time-lapse movies of the Athabasca Glacier in motion, from both above and—beats me how he does it—*below* the ice. His guidebook opens with the story of how Norman Collie, the British chemist who invented the neon lamp, of all things, led an expedition to the summit of Mount Athabasca. The year was 1898. Before him stretched "one of the most spectacular natural features in the Canadian Rockies," a gargantuan snowscape he later named the Columbia Icefield.

> The view that lay before us in the evening light was one that
> does not often fall to the lot of modern mountaineers. A new
> world was spread at our feet: to the westward stretched a vast
> icefield probably never before seen by [the] human eye, and sur-
> rounded by entirely unknown, unnamed, and unclimbed peaks.
> . . . The level snow stretched . . . mile upon mile before us like a
> rolling snow covered prairie. . . . The peaks [rose] . . . here and
> there like rocky islets from a frozen sea.[4]

Roughly translated from the reserved parlance of nineteenth-century
British explorers, this would mean something like: "The scene fairly blew
our woolen socks clean off."

I also learn that had Collie and his crew climbed the same peak three
million years earlier, they would have seen pretty much the same view. Up
there, in the land of eternal snows, the ice age still rules. What keeps this
gigantic white spot on the map is an annual dump of more than 30 meters
(98 feet) of snow. No other place within hundreds of miles around receives
anywhere near this much.

The Columbia Icefield basin acts like a giant catcher's mitt for snow.
The mountain titans surrounding it present the first major obstacle that puts
the squeeze on moisture-laden winds sailing in from the Pacific. The basin's
high altitude guarantees that most of the snow that falls here never melts in
summer—just as during the last ice age.

What happens to all that snow? It piles up, gradually compressing into
firn, a kind of limbo state for crystals in transition from snow to glacial ice.
In its youthful phase, firn is nothing more than a mass of BB-sized blobs of
ice riddled with serpentine air spaces. Under increasing pressure from above,
air within the ice is steadily squeezed out, while meltwater seeps in, eventually freezing and recrystallizing until the deeper layers of firn are transmut-
ed into true glacial ice. This process can take several years, or even decades,
depending on the amount of the annual dump of snow up top. One glaciologist boldly calls the end product a "mono-mineralic rock" since, like quartz
or diamonds, glacial ice is no more than a consolidated mass of a single min-
eral—in this case, frozen water.

As I flip through Kucera's book, my thumb freezes on a full-page
image of what at first looks like a 1960s art experiment in stained glass. The
caption reads: "Photomicrograph of typical crystalline mosaic of glacier ice."

To capture this image, somebody had sliced a microlayer of ice from the belly of a glacier, clamped it onto a prechilled microscope slide, then bathed it in polarized light, revealing an irregular patchwork of rainbow colors. As an ardent snow lover, what really stirs me about this image is that it allows us to peer into the heart of glacial ice and see it for what it really is: not an amorphous hulk, but a frozen concentrate of unique entities echoing the original snow crystals, which drive the whole process of glaciation. I catch myself nodding briskly as I read the words of park interpreter John Wood, who neatly summed this link between heaven and earth: "Life for these glaciers comes from the sky."

So what happens to all that ice? I learn that the center of the Columbia Icefield is as deep as Toronto's CN Tower is tall. This is the glacial spawning bed for the nine valley glaciers that spill from the icefield's brim. What gets them moving is some magic mix of temperature, pressure, weight, and slope, bringing those glaciers down from the mountains like molasses in January. Kucera studied these ice engines back in his laboratory by subjecting small chunks of glacier to the same kinds of stresses and strains found in nature. He discovered that at extreme pressures with temperatures held just below freezing, the internal crystal structure radically altered, allowing him to extrude his ice samples through a narrow metal tube like so much toothpaste.

This ability to deform its internal skeleton—much like a mouse is able to squeeze under a door—gives glacial ice a remarkable plasticity and, in "warm" glaciers like the Athabasca, accounts for about one-third of its locomotion. The greater balance of flow comes from the glacier's undercarriage slipping over its stony bed. Greasing the journey is a thin film of meltwater where ice meets rock.

"My time-lapse films have documented the action of basal sliding near the terminus of the glacier," writes Kucera. "You can see the ice moving and dragging boulders along."[5] Now, *this* I would like to see. Visions fill my head. The glacier's prow plucking out giant boulders, clawing away at mountains, chattering across buried ridges, sculpting and gouging the bedrock with deep crescents and grooves, stripping all life from the land, peeling back its stony skin and grinding it to dust.

I read on, learning that, a million and a half years ago, as the Ice Age took charge of the northern hemisphere, the Athabasca Glacier might have been hundreds of kilometers long and more than 1.5 kilometers thick. En route through the mountains it merged with sister glaciers, creating knife-edged arêtes along the ridges and spiked horns among the peaks. Flowing through the valleys, it gouged armchair basins and U-shaped trenches kilometers wide. As it quit the Rockies and joined forces with continental ice sheets crunching down from the north, nothing escaped its icy blade. Gone were most of the soils, riverbeds, and landscapes of preglacial times. The last glacial advance reached its greatest glory about twenty thousand years ago, capping the continent with 15.5 million square kilometers (6 million square miles) of ice—enough to cover present-day Antarctica. At its thickest, over Hudson Bay, the ice sheet measured almost 5 kilometers (3 miles) from top to bottom. Except for a few pupating insects or dormant seeds trapped in caves beneath the ice, the land was stone dead.

Then, about eighteen thousand years ago, a mysterious force started pulling on the planet's climatic pendulum, triggering a gradual swing back into warmth and renewed life. The ice started shrinking. Like the advance, this process was spasmodic. At times it slowed to a halt, then readvanced, backed up, then readvanced again. Leading this glacial dance was the ice sheet's annual income of snow. When this income exceeded the annual expenditures of melting, the ice sheet advanced. When melting had the upper hand, it retreated. At the best of times, the net pace of glacial retreat was about 90 meters (300 feet) per year—not a bad pace for something so huge.

Though the Ice Age's shadow has long since faded, climatic wobbling continues. As recently as 360 years ago, during the peak of the Little Ice Age, when unusually heavy snows and cold temperatures prevailed worldwide, a glacier tumbling off the slopes of Mount Blanc, France, was said to have advanced "over a musket shot each day, even in the month of August." In 1644, when this glacier threatened to dam the Arve River and flood nearby farms and villages, the bishop of Geneva, with three hundred villagers gathered around him, blessed the toe of the glacier in an act of sincere supplication. The hills, I imagine, rang with the prayer, "May you melt in peace." Had Jasper Park's Icefield Centre been around in those days, it too, gift shop and all, would have been well under ice. Unaided by priestly intervention, the Athabasca Glacier's toe is now parked about 2 kilometers (1.2 miles) away from me and shrinking fast, apparently because we humans are mucking about with the climate.

"The Columbia Icefield may be able to resist some of the ravages of current global warming," writes Kucera.[6] About the Athabasca Glacier he is less optimistic, having watched it waste away at a "spectacular" rate for many years. He figures that since 1870 it has lost no less than two-thirds of its mass. Poof. Gone. Between 1992 and 1994, he observed with his own eyes a glacial retreat of almost 24 meters (80 feet) per year. It seems the Athabasca's snow budget is dipping dangerously into the red. In the grave tones of a man obviously in love with glaciers, Kucera concludes that the future of the Athabasca and its neighboring glaciers looks "ominous."

Jeez, we'd better get out there before there's nothing left, I think, slapping Kucera's book shut. Another paradox: in this so-called land of eternal snow, we seem to be losing our beloved glaciers fast.

I see I have another 10 minutes to kill before liftoff, so I go downstairs to check out Parks Canada's mini-museum on snow and ice. The first thing I walk into is a micromovie, by none other than Professor Richard Kucera, which speeds up the Athabasca Glacier's movement by 25 hundred times. Collapsing 10 hours into 5 seconds, the film shows the glacier's gnarly underbelly slowly grinding a few pebbles to mush. Nothing much to write home about. But elsewhere I read that you could put the Athabasca's slow but stupendous erosive power on par with the detonation of a large nuclear bomb. And all this triggered by a few billion crystals of innocent snow.

Riding Clarence

I feel a tug at my elbow. "Time for the Snocoach, Dad."

As we march up to Gate B, clutching our tickets, I wonder whether they will x-ray my knapsack-full of books before I climb onto the Snocoach. Will they serve us orange juice and pretzels onboard? Will they provide fold-down table trays? You can see I'm new to all this.

After all the hype and the money, what rolls up outside is a plain old bus you might take from Moose Jaw to Medicine Hat. Nothing fancy. No house-sized wheels, no polar bear barricades, no pretzels.

That basic Brewster bus takes us up a special Brewster road to the Brewster Snocoach corral halfway up one side of the Athabasca Glacier (soon to be renamed the Brewster Glacier?). There we meet the giants. Growling behind a row of wooden staging platforms are several Snocoaches, either disgorging or engorging passengers. With massive balloon tires, blocky lines, and goggly headlights, they strike me as ungainly planet rovers, the type that Buck Rogers might have driven over the purple hills of Xenon. Each Snocoach has a different name. Ours is Clarence.

Greg, our driver and guide, endears himself to us with a promise to put Clarence through the paces. "The intense acceleration may cause some passengers to temporarily black out," he warns. "Please ensure that your seatbelts are safely fastened." All the suckers onboard, including me, grope around for nonexistent belts. "Hang on, folks. I'm about to pop the clutch." After a dull thud we pitch, tortoiselike, down a muddy slope so steep I have to clutch the back of my seat to avoid face-planting into a Japanese woman in front of me.

"Bath time," Greg announces at the bottom, as Clarence plows through a meltwater stream at the glacier's edge. He explains how we don't want to track dark, heat-absorbing mud onto the snow and ice because that could make "horrible man-eating ruts."

From within the warmth and safety of Clarence's belly, Greg seems to take glory in pointing out all the perils waiting for those who would dare explore the glacier in any way other than via a Brewster Snocoach. He revels in anecdotes about near-miss avalanches, collapsing ice caves, outburst floods, and falling ice blocks that would flatten a car. In more sober tones, he moves on to "the deadliest hazard of them all"—crevasses, deep cracks in the glacial ice, one of which just last month claimed the life of a Japanese boy. Greg points out several gaping crevasses large enough to swallow my whole family without a trace.

As we rumble over the glacier's corrugated surface, it's time for another mood swing. "How much ice do you think is out there?" asks Greg, now sounding genuinely jolly.

"A hundred football fields," says one American tourist.

"A thousand hockey rinks," offers somebody else, most definitely a Canadian.

"Close," says Greg. "More accurately, there's enough ice out there to make a zillion quizintillion Margaritas." After letting that sink in a bit, he adds, "And guess what. I bet you think this glacier was formed from ice. Right?" Some people glance at each other, pretending not to look dumbfounded. "*Wrong!* This baby is made from one hundred percent snow." He then explains, in refreshingly simple terms, the time-tested recipe for whipping up a glacier. Take *lots* of snow crystals, squish into firn, add more snow, squish again. Allow meltwater to invade spaces between crystals then refreeze mixture into glacial ice. Repeat over many years. Finally, pile on yet more snow, then stand well back. Your glacier's on the move.

"Please enjoy the glacier," Greg says, as he wheels Clarence into a big patch of groomed ice. "But don't go wandering beyond the turnaround area." He looks at his watch. "I can give you 16 minutes for your once-in-a-lifetime trip to the center of the Athabasca Glacier. I'll beep the horn in fifteen."

Then, like a priest before an altar, he stands before a red warning sign over the exit door. "Now, I must ask you to repeat after me: Warning! Persons choosing to disembark the Snocoach . . . " We passengers respond in equally solemn tones. ". . . do so at their own risk. " More mumbles in mixed tongues. On my way out I notice another sign telling me that "warm clothing and appropriate footwear are recommended." Several chipper passengers squeeze past me wearing beach sandals and T-shirts. I wonder if they will still be smiling in 16 minutes.

Outside, what first hits me is the glacier's noise: the murmur of meltwater somewhere beneath my feet, the clatter of pebbles that slump off thawing rubble, the subtle pings and groans of rotting ice. Something is dying here. This whole system seems to be falling apart. I glance up the valley and sample a lungful of puzzling air. The world smells fresh, as if everything could start over—or cave in any second. I can't tell which. A beacon of sunlight breaks through the sinking clouds and ignites the Columbia Icefield's snow-clad brow, now a picture of immortal freshness.

I pull out my official Brewster Snocoach brochure to try and make some sense out of this place. It tells me that here at the turnaround point, halfway up the Athabasca, the glacier is 300 meters (1,000 feet) deep. Three hundred meters of squished snow. The original crystals, way down there, might have fallen from the sky about the time Columbus ran aground off the Bahamas coast. Quizintillions of unique and exquisite crystals now anonymously pasted together in a homogenized hulk of ice.

I learn also that here, just under my wet running shoes, the glacier's surface is suffering a net loss of more than 2.5 centimeters (1 inch) a day. This tired old beast is indeed wasting away. In the same breath I read that there is some eternal secret to be discovered out here on the glacier—"the timelessness of epochs . . . felt by all but the most jaded travelers." Call me degenerate, but after 13 minutes I fail to feel it.

An icy blast of wind knocks the brochure from my hands, and I go running through the puddles after it. Clarence sounds a train-sized horn. I look up to discover that the four of us are the only ones not already onboard. Don't ask for whom the Snocoach beeps; it beeps for thee. We race each other back, with 3 seconds to spare for a family photo in front of Clarence's 1.5-meter (5-foot) -high tires.

On the way back, Greg stops Clarence beside a cavernous, corkscrew-shaped hole in the ice, down which roars a fire hydrant's worth of meltwater. A "millwell," he calls it. These holes sometimes drop all the way to the bottom of the glacier. "Slip in there, and your next stop is the toe of the glacier about a mile downstream," he says. Then he tells us about a rookie Parks Canada guide who accidentally plunged into such a stream, got flushed under the glacier, and popped out from beneath the toe, still alive and kicking. As I imagine myself disappearing down that hole I conclude that such people have horseshoes tucked in the right places.

A few bumps and thuds later, Greg slows down—from maybe 13–8 kilometers (8–5 miles) per hour—to point out a cone-shaped hump of muck out on the ice. "You see that pile?"

We see it, Greg.

"I bet you think it's made out of plain old dirt, don't you?"

No, Greg. Now that you've said this, of course it couldn't be.

"Wrong! It's glacial doo-doo."

He gives it a fancy name that sounds to me like "kryptonite," then goes on, with revved-up enthusiasm, to explain how it consists mostly of dust cores that were once at the heart of billions of snow crystals. I lean forward in my seat. Clarence is making an awful racket as he gears down, and I miss half of what Greg is saying. Something about how this brown mush gets concentrated at the bottom of millwells. Now, *that's* interesting. Like Kucera's stained glass image of the glacier's heart, here was more tangible evidence that this grand glacial show owes everything to snow. What did Greg call that stuff? As if back in grade school, I raise my hand to get his attention. But he moves on to more engaging anecdotes about such things as his climbing exploits, his record of driving offences, and his marital status. After conducting hundreds of tours in this thin mountain air, I suppose one's mind should be expected to ramble a bit. My mind rambles too, back to those odd piles of glacial doo-doo. I crane my neck around, trying to catch one last glimpse of the stuff.

Under the Giant's Toe

Back at the crowded Icefield Centre, I feel like I've just returned from a visit to another planet. Pluto, perhaps. I hear there's a lot of snow and ice way out there. We must put hundreds of miles of pavement behind us today and have just enough time to check out the glacier's toe. But first, I want to visit one of our national parks' most endangered species (thanks to budget cuts): a parks interpreter. Besides, when running for the bus, I spotted a pile of glossy books for sale at the Parks Canada counter. In such matters I have no choice.

While waiting in line for the interpreter—after waiting in line for the washroom—I notice a beverage cooler that dispenses 100 percent pure Brewster Glacier Water. The bottle labels are printed in six languages, including Spanish and Chinese. I didn't check the label on the toilet paper dispenser, but it too might have been genuine Brewster brand. As far as the Brewster empire may reach, this is, after all, a national park and darn it, I want to talk to someone from the institution ultimately responsible for protecting all these glaciers into perpetuity.

My turn finally comes. I suddenly realize that my scribbled but well-formulated questions about snow and ice are in the notebook buried at the bottom of my pack. As this dawns on me, a young woman with 'Chantelle' embossed on her Parks Canada nameplate gives me a steady smile. She's waiting for me to say something, so I ask her the first thing that comes to mind.

"So . . . ah . . . tell me about that Japanese boy who went down the crevasse."

Chantelle shakes her head. "We do all we can to keep people off. You've seen our warning signs all over the place?"

I hadn't actually, though I nod emphatically.

"Still, people wander around out there every single day. Some don't read the signs. Some do but ignore the message. Sometimes people die."

I sense she is politely sizing me up as to which category I might fit.

"They tell themselves, 'Oh, I'll just go up a few more feet'," says Chantelle. "Then, once they get out onto the slick ice, they're often tempted to walk onto the softer snow, not knowing that snow hides the most dangerous areas. Sure, the chance of you falling in may be small. But the chance of getting you out alive is just about zero." She taps the glass case in front of her. "Have you seen this?"

Beside a glorious satellite image of the Columbia Icefield is an *Edmonton Journal* article dated July 14, 1994.

Dr Rainer Bergener, 39, of Germany had hiked up to the toe [of the glacier] Tuesday with his wife and six other German friends. He stepped onto a snow bridge around 2 p.m. and plunged

27

down 10 metres where he was jammed and buried by collapsing snow. . . . Snow builds up in the crevasse over the winter and when it melts, snow left at the top creates a snow bridge. . . . Rescue workers were on the scene within twenty minutes, but it took two hours to extricate Bergener. Rescuers had to use a pneumatic chisel to chop him free of the icy walls.

Although rescue team leader Gerry Ignadson gave the hiker "the best possible chance of survival," Bergener lost consciousness shortly before being lifted to the surface, dying soon after in a helicopter bound for the Jasper hospital. Later, Ignadson had some advice for anyone going on the glaciers: "Although you can hike on them, if you don't know what you're doing, DON'T DO IT!"

Probably looking a bit white, I glance up at Chantelle.

"You were asking me about that Japanese boy?" she says. "It happened on the fourth of July. The boy and his father were out there in shorts and T-shirts, just thirty to forty feet from the toe. The boy tried to hop over a crevasse, a very, very small narrow crevasse. But his foot landed on the slippery cooler side, and down he went. Even our smallest warden couldn't reach him. By the time they pulled the boy out 4 hours later, his body was beginning to freeze solid. More than likely he drowned in all the ice water. It's not the first time we've lost a child to the glacier, and I'm afraid it won't be the last, no matter how many signs we put up."

I shake my head and slip away, without even glancing at Chantelle's stack of glossy books.

✳ ✳ ✳

A flying fist of windborne grit cuffs me in the face the moment I step from my car onto the parking lot at the glacier's toe. Compared to the sea of flawless pavement skirting the Icefield Centre, this lot seems a slapdash affair. Whoever scratched this crude gravel patch out of the chaos of rubble and rivulets all around probably expected it to wash away at any instant. A rusty old backhoe, its tracks engulfed in rushing brown meltwater, dredges a creek just above the parking lot to prevent, I surmise, that very thing from happening. I glance at my car, shiny and quite new, and catch myself wondering whether that creek, if it suddenly misbehaved, could carry it away.

Brenda and the kids explode from the car and begin scrambling up the winding boulder-strewn trail to the glacier. Trod upon by thousands of tourists a day, this trail is said to be unique in all the world. Where else can you jump from your car wearing (despite what all the signs say) shorts and sandals, and, within minutes, be tickling the toe of a gargantuan glacier? It makes a mockery of mountaineering.

The air is cold and dank. It wafts down over a steep rise in rhythmic pants, suggesting the presence of a dragon lurking just over the crest, breathing not fire but ice. Bumptious clouds, the color of lead, crouch in the vast amphitheater of mountains that frame the Athabasca Glacier. We awoke this morning to rain on our tent and fresh snow on the peaks. Summer aside, in a place like this with clouds like that, I'm thinking that our glacial pilgrimage could soon be aborted by a whiteout. But, being the info-junkie that I am, I wave the kids onward, pull up my jacket collar, and make for the Parks Canada signs at the edge of the parking lot. I am especially hungry for more tidbits on those weird dirt piles I saw out on the glacier. On this topic I discover nothing. Instead I am confronted with yet another wall of danger signs.

Three kinds of people read interpretive signs: streakers, strollers, and students. For the ever-restless streakers, the sign may as well not exist for all they look at it. Strollers begin to space out after glancing at a few pictures and maybe reading a caption or two. Students read every word—even who made the sign, if that information's provided. I'm definitely a student, though not an exceptionally brilliant one. Maybe I suffer from tourist-brain syndrome, but I usually have to read interpretive signs a couple of times to get what the heck they're saying. This sometimes drives my daughters crazy, who, like most kids, are streakers. They call me a "sign addict." But on one pass, these signs convey to me a take-home message as cold and clear as Brewster's bottled water: "Those Toying with Suicide Please Step on the Glacier."

What first catches my eye is a yellow triangle showing a human figure falling headlong down a bottomless crevasse. Below are the blood-red words: CREVASSE—DANGER. The text begins innocently enough: "If you've never been here before, you've never been anywhere like this." Then it provides a brief field guide to man-eating crevasses: wide and slippery, snow-covered and narrow, and everything in between. The sign goes on to describe the many gymnastic ways which I might fall in: glissading down a

*A veritable forest of danger signs greets visitors at the Athabasca Glacier. Some peo-
ple get it; some don't.*
Jamie Bastedo

slippery chute, falling through a snow-choked crevasse, busting through a
snow bridge. If that wasn't warning enough, it tells precisely how I might
expire. "You can be seriously injured or killed by impact of the fall. You can
die from hypothermia if trapped in a crevasse. You can drown in meltwater

flowing into the crevasse." Then, in case I didn't catch the drift, it adds: "People have been killed falling into crevasses *on this glacier.*" On that note, the sign ends by politely advising that, "unless you have experience in glacier travel or are accompanied by a licensed guide, travel on the ice is absolutely not recommended."

I glance up the trail. My wife and kids are already out of sight. The darkening clouds move suffocatingly close. I let loose a gasping yawn, ready for some fresher air. But I need a fix of just one more sign, then I'll catch up to them.

Now, *here's* an interesting one. With no pictures or maps and a solid page of text, this sign must be a turnoff to all but the most dedicated student. I bet hardly anyone reads it. But the title hooks me: DANGER—A NOTE TO PARENTS. Then: "Before you lead (or are led by) your children onto this glacier, think about this. . . ." The scenario that unfolds is all too real. I am riveted.

> One moment . . . It's a hot, sunny day. Everybody is in shorts and sandals. It's been a long drive; the kids need to burn off some energy—and you need a break. What better place than a cool, ice-blue glacier to get rid of some of that steam. You race up to the ice, and yes, it's colder up there. Lots of folks are on the ice. Seems harmless enough—let's go up! Soon the kids are running and jumping, having a great time.
>
> You are 15 meters [50 feet] from the edge of the glacier.
>
> The next moment . . . You look back to see your child take a jump across a small gap in the ice. And here's where your life takes a different turn. The ice beside the crevasse is out of the sun and is more slippery. The snow along its edges is weak and rotten from the day's heat. He slips backwards, and without even time for a scream, disappears from view.

What follows is a minute-by-terrifying-minute depiction of what could happen to your child or mine.

> The crevasse is long, narrow, slippery, ice cold and dripping wet. There is no friction to stop his fall, only the gradual narrowing of the crevasse. When he stops he's wedged between walls of ice that hold him fast. . . . You creep to the edge of the crevasse, and

31

hear muffled sounds but see only ice, snow. . . . He's down only four metres [13 feet], but because of his size, it may as well be 400 [1,300 feet].

In this hypothetical scenario—or *is* it? The sign doesn't say—it takes 30 precious minutes for the rescuers to arrive and another 25 for them to set their anchors and descend into the crevasse. By then the boy's muffled cries have stopped. The rescuers find an immobile child covered in snow, sunk into the deepest stages of hypothermia. They can't pull him out. His body heat has fashioned a custom-fit tomb of ice around him.

> The next three hours will see a series of exhausted rescuers raised from the depths of the crevasse with no good news. They will try, desperately, to save your child—for many of them are parents. They will use air-driven chisels, high-tech tools, helicopters, ice-climbing gear, and most of all, a frantic determination. They will fail. And you were fifteen metres [50 feet] from the edge of the glacier. . . .
>
> What you see before you looks safe. It's not. Many people will walk here today. Most days they all come back. Bring *your* children back alive.

What really gets me about this display is that it's signed, "A Parent Who's Seen it All Before."

An icy gust of wind finds its way down my collar and I shiver. My kids should read this, I think. *My kids!* With a knotted stomach, I start trotting up the trail, almost tripping over a low wooden marker with the year 1955 carved in it. This tells me that my car is parked about where the glacier terminated the year I was born. Its terminus—and my family—is now out of sight, over the hill up which I now puff. I pass several family groups coming down from the glacier, chattering in several languages, all members apparently present and accounted for.

I suddenly relax my frantic pace, remembering the caliber of mountain woman escorting my kids toward the glacier. I should tell you that Brenda is among the most cautious people I know, and marrying her has saved my life in more ways than one. Being a western Alberta girl and a veteran of numerous mountain safety courses, she is fully at home among these glaciers and peaks. Around the campfire, my kids and I often prompt her for stories of how she used to glacier-hop on backcountry ski trips, successfully evading more than one avalanche. *She* doesn't need to read those signs.

Climbing easier now, I hear lively chatter fast approaching from behind. A group of hardy hikers, dressed to the nines in yellow and red weatherproof gear, strolls up the hill right past me. They are led by a young mountaineer type who is jauntily twirling an ice ax. As the last hiker passes, it dawns on me that this guy has got to be what the locals call an "Iceman," a licensed guide who reads snow and ice like the rest of us read street signs. Passing right under my nose is an expedition about to do the unthinkable: go for an afternoon ramble over the crevasse-ridden glacier.

Surely this guy could solve the mystery of those strange dirt piles for me. I command my legs to click into overdrive. They balk, but I manage to scramble past the company of explorers and catch up to their leader. Lucky for my legs, he chooses this moment to turn around and address his group. I eavesdrop nearby, puffing into my shirt and trying to look nonchalant. From the brochures back at the Icefield Centre, I know that his dedicated pilgrims paid close to fifty bucks each for an "Icewalk Deluxe." I paid nothing and so slink in the sidelines, hoping to snitch a few crumbs of insight from this glacial guru.

He's in the middle of some profundity, pointing his ice ax downhill at the danger signs I had just read. I take it that he's talking about the heathen crowds who insist on walking all over the glacier.

"It's what I call Darwin at work," he says to his admiring clients. "Natural selection weeds out dumb tourists."

After sharing a few gruesome anecdotes about failed rescue attempts, he resumes his march toward the ice. I notice that his crew now seems bunched even more tightly behind him.

As I watch them pass me by a second time, I decide that it's now or never. What harm is there in picking the Iceman's brains just a little? In such circumstances, info-junkies have nothing to lose. I shuffle up beside him. He greets me with a look that is surprisingly benevolent, given my lowly status. His devotees fall silent.

"Ah, excuse me," I say. "Do you mind if I ask you a couple questions, even though I'm not a paying customer?"

"Go right ahead. My meter starts running once we're on the ice."

I dive right in with reckless abandon. "I'm very curious about those funny-shaped dirt piles out on the glacier. You know, that . . . kryptonite stuff?"

He chuckles. "It's *cryoconite*."

"That would be it. The guy on the Snocoach called it glacial doo-doo. He said something about it being created by the fallout of snow crystals."

"That's one way of looking at it," says the Iceman, and he slackens his pace, as if weighed down by the gravity of what he is about to say. "Those piles are basically made up of anything from the atmosphere that could land on the glacier and be incorporated into the snowcover—volcanic ash from Mexico, dust from Africa, pollen, pollutants, that kind of thing." He points his ice ax at the clouds. "Way up there, ice gloms onto these particles that serve as nuclei for growing snow crystals. A lot of that stuff dropped from the sky, mostly encased in pretty little packages of ice. Besides some dust and algae and junk that blew in off the mountains, what you're seeing is mostly the concentrated atmospheric sludge of billions and billions of snow crystals."

I nod mutely. Though not particularly poetic, his explanation admits a crack of light into my brain. All I can say is, "Thanks. I owe you one." And then I stand aside, at the edge of the trail, watching the expedition press on toward the toe.

In a rare perspicacious pause, it slowly dawns on me that the whole story of this unfathomable mass of ice and snow before me, which shaped a continent and waters it still, began with these cast-off kernels of snow crystals, now clumped into glistening brown piles that look like dinosaur plops. At the heart of all crystals lie these secret chewy centers that get the whole show rolling.

Still welded to my tracks, I find myself nodding to no one in particular as I grasp the cold hard fact that these invisible snow nuclei, made visible in cryoconite, rank among those irreducible nubs of nature—like photons in a shaft of sunlight or neutrons holding a mountain together. Knowing this helps me better understand, in my puny way, the ticking of the universe.

Sheeplike, I fall into line behind the last hiker until we reach a steep crest, beyond which opens an incredible—and I do mean incredible—panorama of the Athabasca Glacier. "Must be seen to be believed," blare the posters in the Icefield Centre. Well, nothing I have experienced so far today prepared my eyes for what they see now. I feel like a jetlagged astronaut stepping from his spacecraft, beholding, for the first time, a planetscape where familiar sensory cues simply do not apply.

For one thing, the glacier's toe is so wide—almost a kilometer across—that I can't look at the whole thing without sweeping my head back and forth in a way that makes me giddy. For another thing, there are these sounds—the gurgling and gushing of meltwater, the groaning and grumbling of flowing ice—all suggesting some breathing, throbbing beast sprawled before me. Yet to look at it, with its inert hulk draped in the skin of a half-decayed elephant, you'd think the glacier was stone dead, like the stoic peaks above it. There is this other sense, too, one disclosed by a palpable swirling in my guts: a sense that among the outpourings of this glacier is an air of quiet menace that cares little if I live. Here, with a cold gritty wind smacking my face and snow clouds sinking fast above me, I feel the solid bottom of my life swiftly fall away, as if I'd just slipped into a crevasse.

I shudder, not so much from the cold as from that feeling a mosquito gets when confronted by a menacing hand. I raise my binoculars to glass around, switching my gaze from glacier to gorbies—tourists like me. I spot Brenda, standing alone with her hands behind her back and her head tipped up toward the mountain tops. We built our Yellowknife home on the flattish "rock plain" of the northern Canadian Shield, so, naturally, Brenda is always happy to "get back in the mountains," as she says.

But where are the girls? Brenda seems distracted by all this mountain glory, perhaps even a bit tipsy, and maybe, just for a rare reckless moment, she's turned her back on the girls. Words from that frightful sign rush back to me even as my binoculars dart from kid to kid: "And here's where your life takes a different turn. . . ." Sure enough, there's a handful of kids blithely scrambling over the glacier's toe as if they were kicking around the neighborhood skating rink. As the sign said, "It seems harmless enough. . . ."

My binos lock onto a familiar flash of green and pink—the jackets of my girls. They are not on the ice, but half *under* it, illicitly catching, in an empty medicine bottle, free drops of glacial meltwater from a suspended tongue of rotting ice. I regain my ability to breathe and close my bowels. I whistle at the girls and wave. They wave back, holding the bottle high.

Through cold ankle-deep muck, I advance toward them, intending to befriend this monster around which they apparently feel so at home. They offer me a sip of their stolen meltwater. "Hmm," I say, after a ceremonial slurp. "Not bad. A bit silty, but I guess it doesn't get much fresher." I don't have the heart to tell them that I detect a strong hint of ibuprofen tablets.

Up close, I discover that the glacier is not so spooky. You might mistake it for a rather overbearing snowdrift that blew in overnight and filled your backyard with crystalline cement. I admire its sculpted riffles and ruts.

I am mesmerized by the deadly crevasses' lithe and lovely contours. Below a gnarly gray skin, their blue-white cores seem to glow from within, reminding me forcibly of fresh-cut whale blubber. I can't take my eyes off them.

We slosh around in the glacier-ground muck, exploring different views of the toe and taking the requisite dozen or so pictures of each other to mark the occasion. "Is that the Iceman?" Jaya asks at one point, as I struggle to reclaim my foot from the muck. I straighten up to look at where she is pointing. There he is, way up the toe. He's stooped over some dark thing on the ice, poking it with his ice ax while expounding to his rapt followers. I twiddle my binoculars, focusing in on the object of their meditation: a big brown blob of . . . what did he call it? Oh yes, cryoconite.

I catch myself imagining America's renowned Snowflake Man, Wilson Bentley, out there on tour with the Iceman. I picture him kneeling beside that venerable pile of cryoconite, sporting his broom-sized moustache and the black wino coat that became his hallmarks. Bentley, for almost fifty winters, painstakingly photographed more than five thousand snow crystals, yet he never saw into the quintessential heart of them as I just had this morning, or at least supposed I had. As smug as I'm feeling right now, I believe that this gentle farmer from Vermont would have moved mountains just to poke around one of those slimy heaps of glacial doo-doo.

For, in truth, what's really inside cryoconite? In that concentrated stew of atmospheric flotsam and jetsam is the protoplasmic germ of life from which all snow crystals are spawned. In those glistening piles are the primordial links in an evolutionary chain leading from dancing flecks of fleece to grinding ice capable of toppling mountains and carving continents. Though it doesn't look like much, cryoconite conceals a magic bridge that spans the seen and the unseen, the tiny and the colossal, the new and the ancient, the eternal and the transient, the giver and taker of life, the beauty and the beast. Can you tell I'm pumped over this stuff?

Without anything up there for crystals to "glom onto," as the Iceman said, what would a planet like ours look like when the mercury drops? Think about it. You might get swirling Jupiter-like clouds, black and bloated with supercooled water vapor that just won't glom. On the other hand, you might get bone-dry, Windex-clean skies with much of your water locked up below, littering the landscape as crude shards of ice. How should I know? But I do know this: you wouldn't get snow. Period.

And, with sincere apologies to Parks Canada, to all those bright-eyed tour guides, and to the entire Brewster clan whose jobs depend on it, no snow means no glacier either—not one among the swarm of glaciers disgorged by the Columbia Icefield, nor the icefield itself. For what is a glacier but a grand sliding casket for the crystal corpses of snow?

The big toe of the Athabasca Glacier, a grand, sliding casket for the crystal corpses of snow.

Jamie Bastedo

And there wouldn't be any nasty crevasses either, come to think of it. In the meantime, somebody else's kids go skipping and laughing over the glacier's toe. I can't watch this anymore. Besides, we've got promises to keep, and miles to go before we camp. I take one last look for the Iceman and his crew. In single file they are marching, as if to heaven, toward an inky cloud that is stealing down the glacier. At our altitude, it's starting to rain. Though the mighty Columbia Icefield is now obscured by clouds, I know that somewhere up there, it's starting to snow. This simple fact makes me smile.

"Let's hit the trail, guys," I shout to my girls, as the lights go down on this dizzying theater of snow.

So, there's no time to sign our lives away and join the Iceman on this vacation. We don't get a chance to poke rented ice axes into that mysterious goo. No regrets. It is enough for me to touch the truth that behind snow's many-splendored paradoxes, behind its purity and power, is an elusive force that spins priceless jewels from dust and ash.

Weeks later, in fact on the very day that snow returns to my Yellowknife roof, I am idly thumbing through Loren Eiseley's *The Immense Journey* when I feel my neck hairs suddenly prickling. He's describing, as only a nature savant like him can, that same cryptic force that creates, dissolves, and recreates snow.

"One can never quite define this secret," writes Eiseley, "but it has something to do, I am sure, with common water. Its substance reaches everywhere; it touches the past and prepares the future; it moves under the poles and wanders thinly in the heights of the air. It can assume forms of exquisite perfection in a snowflake, or strip the living to a single shining bone cast up by the sea."[7]

Chapter 2
Being Snow

Snow is, if nothing else, a cosmic joke, even when the punch line is death. If you lack a sense of humor, you won't understand snow. Frozen precipitation, perhaps, but not snow.

—Bernard Mergen, *Snow in America*[8]

Entering the Treasure House

Shovel through a large pile of scientific musings on snow from the past four centuries and start sifting. You are sure to find it. Buried deep in the forty-odd chapters of the Book of Job is a divine declaration about snow that everyone from scientists to soothsayers is fond of slipping into ruminations on the subject. After grilling Job on exactly *who* gave Orion his belt of stars, commands the eagles, and steers the thunderbolt, God asks him the 64-dollar question about snow.

> Hast thou entered into the treasure house of the snow? Out of whence came the ice? The hoary frost of heaven—who hath gendered it?

Bible thumper or not, I think you can guess how this scene turns out. Job is reduced to a meltwater puddle of humility, admitting that his proud words on such wonders have been "frivolous" at best. "I had better lay my finger on my lips," says Job. "I have spoken once . . . I will not speak again."[9] Job was, according to one religious scholar, "recalled to sane and cheerful trust by the voice of God answering him out of the whirlwind, directing his thought to the wondrous panorama passing before him in the forms and processes of nature"[10]—not the least of which was snow.

Among the many snow inspectors to drop this quote into their writings was A. E. Tutton, back in 1927. I can't tell you his first name. From my reckoning of vintage scientific writing, I, J. D. Bastedo, figure there was some sort of stigma against revealing your John Henry in those days. Anyway, ignoring Job's advice to lay a finger on his lips, so-and-so Tutton wrote one of the first shelf-bending books on the natural history of snow and ice. Tutton's tome is sprinkled with self-portraits that show him posed nonchalantly next to man-eating crevasses or atop cornice-draped mountains. In all the pictures he sports a crumpled top hat and gentleman's cane, apparently borrowed from Charlie Chaplin.

Tutton was the kind of snow scientist who often looked beyond the crystals beneath his magnifying glass to the cosmic hand that made them. You won't find Tutton's evangelical tone in today's scientific literature. This is a shame, really, considering that you're dealing with something so miraculous. Instead you'll read, for instance, that, "Snow is simply"—Did you get that? *Simply!*—"particles of ice formed in a cloud which have grown large enough to fall with a measurable velocity and reach the ground."[11] Or how does this grab you? "Snow is no more than falling solid precipitation excluding ice pellets and hailstones." Or finally and mercifully: "Snow is precipitation of ice crystals, most of which are branched." Now, that's pretty dull stuff. I pulled these quips from an eight-hundred-page handbook on snow which, though excruciatingly informative, doesn't exactly inspire me to get down on my knees and genuflect to the nearest snowbank. On the other hand, Tutton, a snow scientist from more romantic times, likens a snowy morning in Cambridge, Massachusetts, to a foretaste of heaven.

> Perhaps of all the exquisite visions which the author has ever experienced—and there are many—the most abiding is that of one winter's morning in Cambridge . . . On emerging from one's home, in a road lined with fine trees on both sides, linked up by high hedges, the whole avenue was seen to be dressed in spotless white, yet of no ordinary snow or hoarfrost; and every point of it was glittering in the rays of the newly risen sun, as if studded with diamonds as closely as they could be packed alongside each other. On closer examination, every branch, leaf, stem, frond or projection of any kind [was] all carpeted with countless myriads

> of those lovely snow flowerets or delicate ice crystals, every one a marvel of beauty worthy equally of the study of the most gifted savant. . . . The scene . . . was one to be remembered forever as a foretaste of heaven itself.[12]

I shamelessly count myself among the lineage of snow inspectors who, like Tutton, are occasionally stunned to the point of paralysis by one glance at a seismically gorgeous snow crystal. But, unaccustomed as I am to flattery, calling me a "gifted savant" might be pushing it. Nevertheless, thanks to my pilgrimage to the Athabasca Glacier, I have indeed seen with my own eyes what others of my ilk apparently missed: those hearts of stone at the crystal core. I have beheld the mysterious condensation nuclei that raise the curtain on all snow shows—or at least I've caught a glimpse of a big slimy pile of them out on the rotting ice. Many earlier snow pilgrims weren't so fortunate.

Take John Tyndall, for example, British physicist and mountaineer. In 1872 he wrote of our blindness in understanding what triggers crystal growth in the first place.

> It is worth pausing to think what wonderful work is going on in the atmosphere during the formation and descent of every snow-shower—what building power is brought into play! And how imperfect seem the productions of human minds and hands when compared with those formed by the blind forces of nature! But . . . in reality, when we speak thus we are describing our own condition. The blindness is ours; and what we really ought to say, and to confess, is that our powers are absolutely unable to comprehend either the origin or the end of the operations of nature.[13]

Having journeyed to the outer frontiers of snow science—remember, we're talking the 1870s here—Tyndall asks you to "hold before your mind's eye, the unseen architecture which finally produces the visible and beautiful crystals of the snow [made from] particles which no eye has ever seen."[14]

Vermont's beloved Snowflake Man, Wilson Bentley, said much the same thing a few years later, declaring that there is "a mysterious something" around which snow crystals form. He deduced the presence of some kind of nuclei but never had the satisfaction of seeing one. "If such nuclei are, as some assume, dust particles," he wrote in 1923, "they must be exceedingly small, invisible to ordinary microscopic vision, as none are found in my pho-

tomicrographs . . . I have found that science as yet can offer no wholly satis-
factory explanation of these phenomena."[15] In all his five thousand photos
of snow crystals, he spotted not one furtive speck.

*A snow-clad copse of paper birch, the kind of scene that often stopped snow scien-
tist A. E. Tutton in his tracks.*

Tessa Macintosh

History can forgive fellows like Tyndall and Bentley for passing off the nucleation process as heavenly hocus-pocus. That they couldn't explain this process is not surprising. Those flecks of dust and ash at the heart of every snow crystal are rather petite. Today's snow scientists can tell you that the big ones top out at around one micron—roughly the diameter of a flea's nose hair. Most nuclei are about a tenth that size. Thanks to electron microscopes, scientists can tell you that the main source of these nuclei is clay-rich dust kicked up from the earth's surface. Air pollutants come in a close second, with the balance provided by forest fires, volcanoes, bits of plant matter, wisps of sea salt, a carousel of microorganisms, and the occasional flotsam and jetsam that sails in from outer space. They can tell you at what temperature and moisture level cloud droplets, or water vapor, will glom onto these nuclei. They can even predict that in every cubic meter (1.3 cubic yards) of the lower atmosphere, where snow forms, you will find anywhere from several hundred to many thousands of aerosol particles floating around up there, just waiting to kick up a snowstorm.

For today's snow scientists, the nucleation process that gets the whole show going is really no big deal. No divine riddles here. Science has busted this process wide open. Duplicated it, in fact. Now, is that playing God or what?

The first mortal to grow snow from scratch was Japanese nuclear physicist Ukichiro Nakaya, who in 1936 created a minor snowstorm inside his laboratory. Inspired by the snowy moutainscapes of Hokkaido, Japan's northernmost island, and a physicist's zest for the minuscule, he built a special 4 meter-long (13 feet) chamber which could replicate a wide array of atmospheric conditions in which snow crystals might form. With the turn of a tap and flick of a switch, he was able to control the chamber's temperature and moisture conditions with remarkable accuracy–down to -580°F (-500°C) and up to 100% humidity. In spite of his credentials, this was not rocket science. The real trick was to figure out what to use as a growth platform for snow crystals. Nakaya decided to string up various filaments like tiny clotheslines across his chamber; he then pumped it full of superchilled water vapor to see what would happen. He started with minute threads of wool and cotton. All he got were coarse coatings of frost that made the threads look like fuzzy frozen caterpillars. Then he somehow strung up single strands of cobweb. To pull off this stunt, the literature is silent on whether Nakaya employed specially trained spiders, so allow me to start the rumor. All he got, however, was more frozen fuzz. Okay, he thought, how about using fine Japanese silk? Still more fuzz. Then—don't ask me how—Nakaya hit upon the idea of stringing rabbit hairs across his vapor chamber. My reckoning is that when he later peaked through the observation window, all he could say was, *"Sugoi!"* which is more or less the Japanese equivalent of "Eureka!" It turned out that microscopic nodules lining the rabbit hairs served as perfect hatching grounds for the birth and development of bona fide snow crystals—exactly as condensation nuclei do in the clouds.

Equipped with a fresh batch of rabbit hairs, Nakaya started playing around with the temperature and humidity in his vapor chamber. That's when things really got interesting. Storm after mini-storm swept through his laboratory over the next eight years. "By 1944," he wrote in his magnum opus, *Snow Crystals—Natural and Artificial,* "we were able to produce every type of snow crystal artificially, and were therefore able to determine clearly the conditions of their formation."[16] Soon after World War II, Nakaya's accomplishments received an avalanche of attention from meteorologists around the globe, eventually earning him the title of "the greatest snowflake scientist of the twentieth century."[17] Science had faked the snow nucleation process indoors. It was time to try this trick outside. Human evolution had reached the point where we could contemplate ordering up real snowfalls like so much pizza.

They called it cloud seeding. The idea was to knock the stuffing out of constipated clouds by injecting them with a laxative of extra nuclei. The first successful cloud-seeding experiment was conducted in 1946 by one of Nakaya's biggest fans, American meteorologist Vincent Schaefer. Instead of aiming a planeload of rabbit hairs at an unsuspecting cloud—and you have to wonder if the idea ever occurred to Vince—he let go a shower of solid carbon dioxide, or dry ice, into a supercooled cloud. Imagine the smile on his face when he observed the sudden proliferation of snow crystals outside his plane window. And get this: minutes later the cloud he'd just flown through vanished, presumably having dumped its load.

Not to be upstaged by Schaefer's magic tricks and concerned about the threat of weather warfare, the US military launched Project Cirrus over the cloudlands of the American Southwest. Cloaked in secrecy from 1947 to 1953, this was the first operational test of seeding clouds with silver iodide. After more than one hundred foiled attempts, they managed to shake a modest storm of snow crystals from a supercooled stratus cloud by injecting it with pieces of burning charcoal impregnated with silver iodide. Now, just why burning charcoal was employed to help trigger a snowfall, the report did not say. Nor did it say whether the success of this grand experiment was tempered by the ignition of forests or ski chalets far below.

Fiddling with clouds to dump snow when and where you want it soon became something of an obsession among meteorologists and commercial interests whose bank balance often varied with the weather. Ski resort operators, farmers, hydroelectric companies, foresters, airport managers, and a small army of dog mushers all had a stake in this business of snowmaking. A flood of US tax dollars poured into universities and research institutions for weather modification studies, many of them earmarked for "snow augmentation projects." Seeding experiments continued, often with close scrutiny from the media. Headline stories sparked public debate on the feasibility of regulating North America's weather, if not the world's. The sky seemed no longer the limit.

Trouble is, the verdict is still out on whether seeding clouds to harvest snow actually works. In 1957 a weather advisory board lobbying the US government for yet more bucks reported that seeding had squeezed 10 to 15 percent more snow from clouds over the western mountains. Giddy with their presumed, though unproven, success, board members viewed themselves as tomorrow's captains of weather technology, controlling the rate and form of snow crystal growth, accurately targeting the placement and volume of snowfalls, and eventually—stay tuned for this—managing global weather patterns by thermonucuclear heat. Designer snow was just around the corner. Twenty years later, the same board, now scrounging for dough, lamely promised that "extensive cloud seeding will soon be a proven technology."[18] About the same time, *Science* magazine debunked the whole idea in an article whose title pretty much says it all: "Cloud Seeding: One Success in 35 Years."[19]

If current federal budgets to support cloud seeding for increased snowfall are any indication, this practice may soon go the way of the zeppelin. Besides, in a 1975 government poll of northern Californians, half the respondents endorsed the conventional wisdom that we should simply take the weather as it comes. "Don't tinker with it," they said. "You might upset Nature's applecart." Almost 40 percent went even further, declaring that cloud seeding violated God's plans.[20]

Still, whether it works or not, even whether it's sinful or not, over half a century of cloud seeding research and experimentation has made snow scientists a whole lot wiser about how a few gazillion paltry flecks of dust and ash can fertilize a thumping good flurry. But exactly why their crystalline offspring display such staggering symmetry and diversity remains no less a puzzle than it was back in Job's day. If anyone tells you they have all this figured out, smile politely and know that you are being lied to.

Growing Riddles in the Sky

Chet Raymo, a Massachusetts physicist, has given about as much thought to how snow crystals grow as any man could. When it comes to wrestling all the mysteries out of snow, Raymo adopts a healthy skepticism for his own profession. In his sparkling collection of meditations on science and nature, *Honey From Stone,* he declares that: "Physicists are content that they can explain the hexagonal symmetry of crystals, but they can say very little about the delicacy of the branching and the extraordinary congruity of the six points. For these things, science has provided only the beginning of an understanding."[21]

Some of Raymo's cronies once tried to simulate the growth of snow crystals through computer modeling. They crunched in what numbers they could, reflecting their limited knowledge of the molecular energetics of growing crystals, then stood back. What appeared on the computer screen were crystal-*like* graphics that did indeed display the hexagonal symmetry that underlies the six-sided architecture of water molecules. One stroke for science. But their designs were blocky, crude, and imperfect. Complete artistic flops compared to the real McCoy.

My bet is that no amount of trips back to the drawing board could ever get a mere computer to spit out anything but bogus-looking crystals. What's missing in their model is some kind of organizing formula that explains how snow crystals grow with such utter perfection. How, for instance, do molecules on one side of a growing crystal "know" what position and at what rate they should arrange themselves to match exactly what's happening on the opposite side? This is quite a trick when you consider that, at the molecular level, opposing arms of a growing crystal are light years apart.

In the creation of real snow crystals, some force as yet beyond the grasp of science is in charge of the 3-D artwork. Raymo calls it a "cooperative phenomenon" that simultaneously keeps in touch with all points of the growing crystal. Here's where things begin to get really spooky.

> Some physicists think that *vibrations* of the crystalline lattice are the instrument of communication, vibrations that are exquisitely sensitive to the shape of the crystal. If this is so, then the growing snowflake maintains its symmetry in the same way that members of an orchestra stay in consonance, by sharing the sound of the ensemble. The snowflake's beauty, then, is orchestral![22]

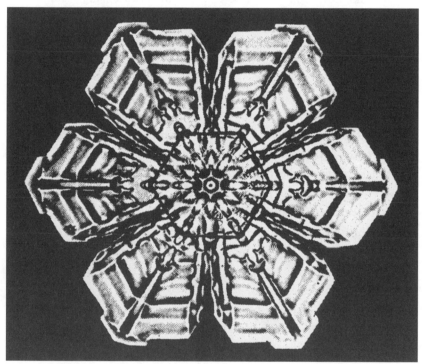

The celestial software that creates such flawless gems remains beyond the grasp of science.

Wilson Bentley

Raymo added the exclamation mark, not me. He has reason to be rhapsodic. His kind of people take pride in tackling subatomic conundrums head-on. To date, all they have discovered at the heart of a growing snow crystal are more questions. "The mystery remains," says Raymo, "—all abiding, undepleted, inexplicable." What I like about Dr. Raymo is that this doesn't bother him in the slightest. He embraces the mystery with open arms, greet-

ing every crystal upon his sleeve as a welcome "cipher." This physicist-philoso-
pher gladly admits that there are fathomless forces at play whenever a crystal
is born. "I can tell you the number of molecules in each flake of snow," he
states matter-of-factly. In the next breath, he humbly adds, "I know nothing."[23]

Unsolved riddles about how crystals grow have not stopped snow scientists from pigeon holing the end products into neat and tidy categories.
My impression is that imposing order on nature's extravagant creativity is
something of a burning itch among snow scientists. My *Handbook on Snow*
basically asks, how could they resist? "It is difficult to imagine such a classification system not being created since man, when faced with the inherent
beauty and variety of different types of snow, would try to communicate this
information with his fellow men."[24]

One of the first snow scientists to publicly admit this compulsion was
nineteenth-century Scotsman William Scoresby. According to one weather
historian, this son of a sea captain, "did as much for snow crystal science as
all previous workers combined." Scoresby cut his academic teeth on physics
and chemistry at Edinburgh University. But on his 21st birthday, he jumped
the ivy league ship to become skipper of a whaling clipper bound for the
Arctic. He spent much of the next ten years plying polar seas, taking time
out from slaughtering whales to study the snow crystals landing on the
sleeves of his brass-buttoned uniform. In his 1820 *Account of the Arctic Regions,*
Scoresby published 96 impeccable drawings of snow crystals, which include
most of the common and not so common forms that you might catch on
your own sleeve (brass buttons not required). Among them were several
"very rare" twelve-sided crystals whose existence has since been legitimized
in the annals of modern snow science.

Scoresby was a consummate lumper. His arctic narrative included a
snow classification system that dumped "an endless variety" of crystal types
into five handy categories. But within each category, he was clearly blown
away by the immeasurable variations on a theme. Turning his back on science, he could only explain this riddle as whimsical slights of the Creator's
hand. "The particular and endless modifications of similar classes of crystals,"
he wrote, "can only be referred to the will and pleasure of the Great Cause,
whose works, even the most minute and evanescent, and in regions the most
removed from human observation, are altogether admirable."[25]

That burning itch to describe and classify the full gamut of crystal
types has driven more than one snow scientist to great heights, sometimes at

their peril. So discovered British astronomer James Glaisher, who moon-lighted as a meteorologist with a passion for studying snow crystals at rather high altitudes. Glaisher was intensely curious about how different cloud temperatures, winds, and moisture levels affected the shape of snow crystals. In the 1860s, long before the Wright brothers were even born, this line of work was challenging, to say the least. Not content to gaze speculatively at winter clouds scudding above his office window, Glaisher took up balloon-ing. He actually became quite good at it.

Glaisher and an unnamed assistant—probably a keen but dreadfully naive student, if I remember graduate school correctly—became adept at ascending into snowstorms. They survived several missions, penetrating cloud after swirling cloud pregnant with snow, while gently swatting crystals with a dark fine-meshed net. The idea was to examine different crystal types formed at various altitudes and atmospheric conditions. Swinging through the clouds, Glaisher discovered Job's treasure house of snow, collecting a kaleidoscope of virginal crystal forms that "gorgeously combined laminae and prisms in the richest profusion."[26]

Exquisite hexagonal plate with a lotus-shaped heart.
Wilson Bentley

We can thank Glaisher for making the first scientific inroads to under-standing the conditions in which different snow crystals form. For instance, he documented the presence of hexagonal plates in dry air at about −15°C (5°F). In clouds with similar humidity but temperatures ten degrees colder, he discovered long column-shaped stars. In warmer clouds with high humidity, the classic Christmas card shapes filled Glaisher's net. Struck giddy by his soaring success and the insatiable lust of a treasure hunter, Glaisher one day pulled all stops, gassing his balloon up to a bracing 8800 meters (29,000 feet). Had he been chasing snow squalls over Nepal, he might well have crashed into the summit of Mount Everest. That's how high the guy was sail-ing, for Pete's sake.

Now, as much as snow may jingle your bells, this is going a bit too far, don't you think? How quickly this thought dawned on Glaisher, we'll never know, but at some point his passion for swatting snow crystals gave way to an overwhelming desire for a late afternoon nap—induced by anoxia of the brain. Without so much as saying, "Night-night," to his assistant, he collapsed in a corner of their wildly swinging basket and passed out. Glaisher's assistant was next to useless, his freezing hands having turned to blue cement. He even-tually saved the mission—their last, by the way—by pulling a cord with his teeth to release gas from the balloon causing a hasty retreat from the clouds.

Back on firm ground, Glaisher published more than 150 amazingly detailed drawings of crystals—especially amazing when you realize that he sketched many of them using only a magnifying glass while on the fly. "Their forms were so varied," he wrote, "that it seemed scarcely possible for con-tinuous observations to exhaust them all." Perhaps overwhelmed by what he called an "infinity of snow crystals," Glaisher, in the end, made no serious attempts to impose order on the potpourri of shapes that filled his note-books. He decided instead to "defer all conclusions for the present."[27] Reflecting on Glaisher's high-flying misadventures, another meteorologist later wrote, "Until such ascents are made systematically and with a careful record of flake forms, for which an elaborate classification of snow will be necessary, our knowledge will remain small."[28]

About 130 years later, snow scientists were still taking to the skies, dar-ing to unlock the treasure house of snow. Part of this quest was to maybe, just maybe, find two *identical* crystals. Nancy Knight, a snow scientist for the National Center for Atmospheric Research in Boulder, Colorado, claims she's done it . . . well, sort of. In November 1986, she chartered a research plane to take her up to a mere 6000 meters (20,000 feet); Glaisher had her beat by almost three thousand, but look what happened to him. Instead of collecting with nets or coatsleeves, which can do a real number on snow crystals when you are traveling at 190 kilometers (120 miles) per hour, Knight used special high-speed cameras. After several months of painstakingly describing and cataloging the thousands of strobe-lit crystals captured on film, Knight had an important announcement to make to the world at large. Among her collection were two column crystals with vase-shaped hollow cores. Rub her eyes as she might, she could find no differences between them. They were "virtually identical," she told her skeptical peers.[29]

When pressed on this point, she admitted that they were, well, "if not identical, certainly very much alike." Pressed further still, she admitted that her monumental discovery hung on only one photo taken from one angle at one magnification. And so you can relax. Knight's sort-of-discovery still leaves the door wide open for endless scientific banter on the no-two-alike debate.

While science can nonchalantly cut and splice genes like so many words on a page, or fire digital images of my kids' artwork to grandparents a continent away, it still can't figure out exactly what the heck's going on up there when it snows. The head scratching continues, even as 21st-century technology takes aim at innocent crystals with new snow-catching instruments worthy of at least a B-grade sci-fi movie. Micro-radar scanners, parallel laser beams, stereoscopic video cameras, supersensitive scales detecting weights as low as 0.00992 grams (0.00035 ounces) (better not sneeze on it)—all kinds of gizmos have generated reams of fancy data and well-educated guesses about the formation and fall of snow.

As sexy as these new tools may be, many scientists will tell you secretly that they have made us little wiser. In a quizzical mix of erudition and caution, one told snow historian Bernard Mergen that such high-tech toys require "more knowledge of the system being studied." Forgive me, but is this not the egg trying to blame the chicken for poor results? What he's really saying is this: while our technology to study snow crystals continues to expand in frankly astounding ways, our understanding of their outrageous creativity lags far behind.

When you think about it, the probability of finding two snow crystals exactly alike is astronomically lower than that of you bumping into one of six billion humans who has your particular nose, your mélange of teeth, and that secret birthmark you tuck away for special occasions. Consider the number of snow crystals that have fallen on our planet since the first flurry, say, four billion years ago. Don't ask me how, but some physicist came up with an estimate of 10^{35} crystals. Sorry, I don't have an "-illion" name for you, but to get a visceral sense of this, grab a pencil and a wide piece of paper and write the number one followed by 35 zeros. How much does that amount weigh? Okay, first imagine catching one snow crystal on the tip of your tongue. At about a millionth of a gram, you barely felt it, right? That's about ten million crystals to the cup. Now imagine the weight of the earth . . . are you feeling it? You can close your eyes if you like; no one is watching. Now multiply the earth's weight by fifty. On your shoulders is all the snow that's ever fallen on Earth. That's one mega-snowball.

Surely, somewhere in all that snow, two crystals *must* have been alike. But consider this. Each crystal contains something like 10^{18} molecules of water—that's one quintillion (unlike gazillion, I did not make up this number). Each molecule can be arranged in lots of different ways. How many? As each crystal gradually grows in the clouds, its quintillion water molecules flip back and forth about a million times until they finally settle down and lock into position. All this molecular scurrying about opens up an incredibly huge number of possibilities for constructing a six-sided crystal. If you really want a number, the best I can do is to say that this is trillions and trillions of times greater than the total number of crystals that has ever fallen.

If that isn't enough for you, tell me, what are the chances that two crystals could have identical histories? I mean, could they possibly drift through *exactly* the same atmospheric conditions—temperature, humidity, winds, and all that—which influence the rate and nature of their growth? Don't bet on it. If you refuse to believe me, look at the identical twins down your street. One wears pigtails, the other has a buzz-cut. Their personal histories set them miles apart. Same story for snow crystals.

So again, what are the odds of finding two alike? About as close to zilch as you can get. My humble statistics dash all hopes of ever finding them. Shall we end it there?

Snowflake Man Wilson Bentley didn't need statistics to arrive at the same conclusion. He knew it in his guts after four decades of swooning over snow with his camera. You can take his word for it.

New and beautiful designs seem to be as numerous now as when
I began the work forty years ago. While many of them are very

similar one to another, I have as yet found no exact duplicates. In this inexhaustible storehouse of crystal treasures, what a delight is in store for all future lovers of snowflakes ... Oh for a thousand hands, a thousand cameras, to preserve more of this exquisite beauty so lavishly scattered over the earth. And yet there should be no despair, for this miracle ... will come and come again for all time, either here or somewhere in the universe, for beauty and life and love are eternal, the things that make the universe worthwhile and justify its existence.[30]

In search of the miraculous. Wilson Bentley photographing snow crystals, Vermont, ca. 1917.

Wilson Bentley

Redeeming Plain Old Snow

Scratch the surface of published snow nomenclature and you'll discover that something odd happens once those crystals hit the dirt. You leave behind such endearing terms as diamond dust, Christmas stars, and tom-toms only to encounter a host of ignoble descriptors like globules, grains, and crud. Those virginal shapes that fall from the heavens by the trillions—plates, columns, cups, needles, scrolls, and stars—all start decaying, shrinking, atomizing, or otherwise self-destructing the moment they land. That's with or without the help of wind, sun, or your thrashing ski poles.

Don't get me wrong. That doesn't mean the creativity ends here. But to some who would sport archetypical snow crystals on their ties or ear rings, there's nothing much to get excited about once snow has fallen. Even some snow scientists fall into this category of disdainers, says Bernard Mergen in his must-read book, *Snow in America*.

> For many people falling snow is more beautiful and fascinating than fallen, the accumulation more intriguing than ablation [melting]. I have listened to meteorologists who were eloquent on the subject of snow formation but indifferent to snow on the ground. For them the life of the snowflake ended when it reached the earth. By contrast, for glaciologists and hydrologists the snowflake is prenatal. They seldom look up. Their snow is born in the crust of the fallen snow and dies when the last crystal melts back into water or evaporates into the air. From air to solid to water is the life cycle of snow, and its worship requires a true trinitarian creed. Solid, liquid, and vaporous forms of H_2O—each deserves its exegis.[31]

Then there are those who harbor varying degrees of aversion to snow in *any* form, falling or fallen. So I inferred about CBC broadcaster Arthur Black when he once introduced me to a national radio audience as "a snow nut." Keep in mind that Arthur lives in the Canadian subtropics of Salt Spring Island, off the steamy southern coast of British Columbia. He came to my home town of Yellowknife a couple springs ago as a celebrity host for the Arctic Winter Games. That's when I converted him. To a love of snow, I mean.

Arthur's CBC radio show, "Basic Black," ran for almost twenty years and was billed as "an eclectic blend of unusual and unexpected things in everyday life. It celebrated the offbeat side of people and events." Somebody tipped him off about my kooky passion for snow, and I guess I fit the bill.

Our broadcast rendezvous site was an open jack pine forest on the wilderness fringe of town. To beef up the acoustic atmosphere—for how much noise can you squeeze from a snowdrift?—and to make Arthur feel more at home, I built a roaring campfire and put on some blueberry bannock and tea. Conditions were perfect for thumping my snow gospel: a dazzling spring sun overhead, not a breath of wind in the air, and a fresh fallen blanket of glittering subarctic snow—the fluffiest on earth. Surely Arthur would come to understand that the merits of snow did not end with skiing and igloos.

I knew I had my work cut out for me when he began the interview by describing me as "a man who thinks *plain old snow* is the greatest treasure on earth."[32] Then he let fly one of his rapid-fire, Bugs Bunny laughs. "They told me to meet you here to talk about snow. Okay, as far as I can tell, it's white, it's cold, it lies around for awhile, then it melts. What else can you say?"

"That's what a lot of us do," I said. "Because snow is pretty much in our faces for so long, we tend to take it for granted. The reaction of many southerners to snow varies from a glum acceptance to near hysteria when it falls. I know. I once lived in Vancouver."

"It's true," Arthur admitted. "When it snows, most people see something they have to shovel or sweep off their cars."

Here Arthur had hit a sore spot for me. "I confess, Arthur, that while scraping snow off my windshield, I've actually been late for work just staring at the snow. It's truly painful for me to wreck those wonderful crystals."

Arthur raised his eyebrows to great heights, then declared, "Jamie, if you can talk that way after a long Canadian winter, you're a sick person."

Now, I can understand how someone from Salt Spring Island might come to that conclusion. But I knew where the treasures lay and how to mine their riches. I reached over to a snowdrift and pulled out my favorite tool for examining snow, prechilled and ready for action.

"Oh, you hid a frying pan in the snow!"

"Arthur, this is not a frying pan. This is my crystal catcher."

"Forgive me. It looks a lot like a frying pan to me."

"Well, indeed I have on occasion used it to make an omelet or two. I have another tool down here somewhere. . . ."

"A spatula!"

"Arthur, this is my snow scalpel."

"A snow scalpel?!"

"Correct. These are my main tools for studying snow."

With a note of concern in his voice, Arthur asked me, "Jamie, how long have you been out here in the snow? Do you ever hear voices?"

"I'm really okay, Arthur." I pulled a third tool from my pack. "These are binoculars, right?"

"Knowing you, probably not."

"Right. This is, in fact, my crystal scope. You turn the thing upside down—"

"And you look at snow through the wrong end?"

"Well, sort of, Arthur. For *bird* watchers it's the wrong end. For *snow*-watchers it's the right end."

"Of course!"

I steered Arthur over to a patch of deep, fluffy, virgin snow. "Time for a snow dissection. Let's have a look at the different layers. . . ." The surface snow was relatively fresh, just a day or two old. With a few gentle passes of my snow scalpel, I skimmed a few hundred crystals into my crystal catcher and spread them around. "You've got to prechill this pan—I mean, er, this crystal catcher—or else all you end up with is stir-fried snow crystals." Arthur zoomed in with the crystal scope. "Just look at them, Arthur. Aren't they pretty?"

"Oh yeah. . . . They twinkle just like diamonds!"

"These are classic stellar crystals, like the ones you see hanging in windows at Christmastime."

"Beautiful! I've never really looked at them."

Our dissection—and Arthur's conversion—continued. We examined the old snow lying within the dense middle layers. I explained how, regardless of original shape or life history, all crystals eventually are reduced to that shape nature loves so much: a sphere. "It's kind of sad, Arthur. All that beauty converted to an undifferentiated blob of ice."

"Your problem is you haven't shoveled enough driveways, Jamie!"

We dug down to rock bottom. "I've saved the best till last, Arthur." I showed him *pukak,* the large chunky crystals, sometimes called "depth hoar," found at the very base of a mature snowcover. *Pukak* is an aboriginal term from northwest Alaska; the street term is "sugar snow."

More oohs and ahs from Arthur. "They're fantastic! They look like broken glass."

"Exactly. They remind me of miniature Japanese pagodas or log cabins carved from glass."

During that 20-minute interview, we dug up a lot of snow, mangling a few million crystals in the name of public education. Eventually Arthur's newfound ardor for snow gave way to his persistent addiction to warmth. "How about toasting to all those wonderful snowflakes that gave up their lives for our show today?" he said, sidling up to my campfire.

"Okay, Arthur. Let's do it again sometime."

"You betcha." Arthur let go a final belly laugh, then gulped down some steaming Labrador tea. "Maybe next summer!"

As Arthur Black discovered, snow on the ground is much more than a bunch of sorry-looking skeletons of fallen crystals. One snow scientist with an unabashed fondness for both worlds of snow—falling and fallen—was G. Seligman, who, in his 1936 *Snow Structure and Ski Fields,* asserted that "the life story" of a snow crystal does not end with the journey from cloud to earth. "It has many other stages to pass through before it once more resumes its less exciting role of a drop of water."[33] Granted, you might shove your mitt into a pile of plain old snow and come up with nothing but gnarly grains and globules. But when you think of their bewitching life story, you too have to fall for them. As George Orwell might have said, "All snow crystals are equal, but some are more equal than others." Listen to the heartfelt equanimity of snow statesman A. E. Tutton on this point. In homage to the many faces of fallen snow, he reverently declared that some were "grotesque, others forbidding and dangerous, yet again others wonderfully beautiful, and *all* deeply interesting."[34]

The metamorphic forces of heat, moisture, compaction, evaporation, sublimation, and the sovereign hand of entropy all mold snow on the ground into shapes as arresting and eclectic as any freshly minted crystals that might land on your sleeve. Science's attempts to corner this complexity with names and categories and subcategories and sub-subcategories have resulted in what one snow scholar calls "a certain amount of indefiniteness in the nomenclature." Some of the names kill me. I recommend snapping your fingers as you rhyme these off.

> Sand snow, wild snow, crud snow, penitent snow;
> Ballroom, boilerplate, buffy snow, mashed potatoes;
> *Korn schnee, skavler, uusi-lumi, sastrugi;*
> Snowphistry, snow turds, pseudosnow and snirt.

Bernard Mergen is a man who really digs snow.

> A pit dug in the snow, like an archaeologist's trench, reveals the
> history of the snowcover. Near the bottom of the pit are the

fragments of snow crystals that began forming months before and thousands of metres above the earth. Metamorphosed beyond recognition, these snow grains are to the original snow crystals what butterflies are to caterpillars.[35]

A pyramid of pukak, *ice-ribbed butterflies of the snow world.*
Wilson Bentley

What Mergen is talking about here, of course, is pukak, which, more than any other kind of snow, has stolen my heart. Pukak, that fragile layer of large loosely spaced crystals at the bottom of the snowcover. If you're lucky enough to live in a place where the snow gets half a chance to ripen, instead of melting on touchdown or getting run over by a truck, you'll find it. I even write love letters to it. Here's a quip from my snow journal, written after an elevating December walk on the lake ice behind my home.

A walk on Back Bay to mail my daughters' Christmas cards out east. I give in to a sudden urge to lie prone on the neighborhood skating rink for a few moments to check out the latest offering of snow. I study a thin layer of fresh fluffy snow with my grandfather's hand lens. Delicate stellar crystals lie above irregular needles and rime-rich grains. No one has shoveled this rink for a while, and I happily discover a rudimentary layer of pukak crystals growing below. Ah, pukak . . . so lovely to see you again, growing in perfect angled arches against the thick black ice.

Five months later, in early May, I'm poking around the poplar woods behind my cabin, saying good-bye to a rotting patch of snow forsaken by winter's retreat. For old time's sake, I dig through it with my jackknife (no spatula in my pack that day) and discover some valiant pukak crystals hanging on for dear life.

These last crystal veterans are all that's left of this year's snowcover that once reached from here to the Mississippi valley. They are remarkably intact, furnishing a flawless crystal palace, complete with domes and turrets, for some waking ant or beetle who might stumble into this dying patch of snow.

I've read up enough on pukak crystals to get a good inkling on the *processes* that form them—sublimation, recrystallization, that sort of thing. I'll get back to you later on this. But what I'm getting at here is that, dig as I might, nowhere have I found a reasonably convincing explanation of the *product*. Why do rotten old ice grains in a snowdrift's armpit reincarnate as exquisite ice-ribbed butterflies of the snow world? Apparently no one knows. Though this question hasn't received much air time in the scientific press, the forces that create the marvelous architecture of pukak crystals are as fathomless as those that fashion crystals in the clouds. And frankly, my reading is that it's the same for much else that is going on within the snowcover—at least at the cause-effect level, where I like to snoop around.

According to one handbook for mountaineers, who must know how to read snow to stay alive, "the repertoire of snow is too large for all its tricks to be encountered in one year—or in ten. Complete familiarity with the behaviour of snow, if attainable, would require a lifetime."[36] Call them "tricks" if you like. Loren Eiseley went further, chalking snow's mysteries up to "those obscure tricklings and movings that stir inanimate things." Tricks or tricklings, I personally think that snow has more up its sleeve than we *can* think. That's one of the reasons I'm so nuts about it.

And so I give the last word to physicist Chet Raymo, a fellow snow nut who thrives on nature's unsolved riddles.

> Has the mystery of the snowflake, then, been entirely plumbed? Certainly not! . . . It is the most common of misconceptions about science that it is somehow inimical to mystery, that it grows at the expense of mystery and intrudes with its brash certitudes upon the space of God. . . . In a world described by science, mystery abides in the space between the stars and in the interstices of snow.[37]

By Job, I do believe that Dr. Raymo has got a point there. Call it a "foretaste of heaven" if you like, but there is definitely a mysterious something we just can't get our scientific heads around when you start to appreciate snow just being snow.

Odyssey of a Snow Crystal

Go with me on this—a winter's tale about the life of one snow crystal just doing its thing. Like Snowflake Man Wilson Bentley once said, "Was ever life history written in more dainty hieroglyphics?"[38]

❈ ❈ ❈

Defying gravity, a microscopic fleck of volcanic ash sailed freely, 12 kilometers (7.5 miles) above a sawtooth chain of limestone peaks. Ten weeks had passed since its ejection from the bowels of a simmering Filipino volcano on the other side of the planet. The upper atmosphere was relatively calm this November full-moon night, and temperatures had stabilized at −25°C (-13°F). All around but not touching the fleck of ash were invisible molecules of water vapor, oscillating randomly through space, free from each other's attraction.

Moisture levels in the air rose gradually through the night as warm westerly surface winds swept over the snowy mountain peaks and sent more water vapor aloft. Slowly, the upper atmosphere became supersaturated with vapor. The inevitable collision occurred just before dawn as the moon sank toward the horizon. A lone molecule of water bumped into the ash fleck and instantly condensed into a seed crystal of solid ice. One by one, other molecules of water vapor were magnetically pulled toward the growing crystal and fused to its rapidly growing skeleton. Within each molecule, two small hydrogen atoms aligned themselves around a larger oxygen, much like the ears on a Mickey Mouse hat. Each hydrogen atom stuck out at an angle of 120 degrees, giving the growing crystal a six-sided symmetry that echoed the molecular shape of water.

Column crystal capped with hexagonal plates.
Wilson Bentley

Within 2 hours the crystal encasing the ash fleck had increased its size a thousand-fold. Like the tens of billions of snow crystals now hovering nearby, it had assumed the shape of a perfect hexagonal shaft. Together these pencil-shaped columns of ice drew a thin milky sheet of cirrostratus clouds across the eastern sky. When rays from the rising sun shone through the clouds, they broke into a wide halo of orange and green light—a sundog.

Far to the west, in the foothills of the Mackenzie Mountains, a native trapper emerged from his log cabin to gather firewood for his stove. While reaching for a birch log high on his woodpile, he happened to glance up at the brilliant ring around the sun. He remembered how his grandfather had told him that a sundog meant that the sun was pulling its parka ruff up around its face. "It's going to snow soon for sure," his grandfather would say whenever he saw that ring of light in the sky.

The pencil-shaped snow crystal continued to lengthen, molecule by molecule, until it became too heavy for the faint updrafts to keep it aloft. It started to fall toward the mountains far below. Just before leaving the ice cloud, it collided with two other crystals: perfect six-sided plates which glued themselves to each end of the column. As a whole, the crystal now looked like a free-falling Roman pillar, complete with pedestal and lintel. This capped column drifted slowly through uniform atmospheric conditions until it crossed an abrupt threshold 8 kilometers (5 miles) above the earth. At this elevation the air suddenly became much wetter and more turbulent. Swirling convection currents sent the crystal tumbling rapidly, end over end. One plate disintegrated into splinters of ice. Part of the other plate tore free and sailed off into the night. The rigid column structure endured despite strong mechanical stresses created by the wind.

Moments later, the spinning, half-capped crystal penetrated a water cloud. It was composed of supercooled droplets held in fragile suspension fifteen degrees below the temperature at which water normally freezes. The end of the crystal with the broken plate acted as a kind of magnetic screen, attracting water droplets which froze instantly into tiny uneven granules of rime. As more ice built up on one end, the crystal gradually stopped tumbling and dropped, rimed side down, through the cloud and out into the clear mountain air below.

Four kilometers (2.5 miles) from the ground, the column crystal crashed into a large six-armed stellar crystal, which was falling at a much slower rate. The two interlocked crystals descended steadily for another 3 kilometers (1.9 miles). Surface winds increased dramatically beyond this point, and the crystals were swept sideways for a few minutes, then suddenly upward, all the way back into the water cloud. The resulting wind stress and heavy riming took a heavy toll on the stellar crystal, which was reduced to a two-armed, ice-encrusted vestige of its former splendor. The attached column retained most of its structural integrity, although it too was heavily coated with grains of rime during repeated journeys in and out of the water cloud.

When the crystal finally settled to earth 30 minutes later, only the rigid column structure remained. The stellar fragment had been knocked off during a collision with a hailstone. What was left of the attached plate had become so thick with frozen rime that it snapped off from centrifugal forces alone, leaving one end of the column with a jagged, irregular edge. Still encased inside the column remnant was the ash nucleus around which the original crystal had formed. Near the Yukon–Northwest Territories border, this primal alliance of ash and ice now rested on a steep south-facing slope below a fish-tailed peak with no name.

It had taken 3 hours for the snow crystal to make the 12 kilometer (7.5-mile) journey to this spot. The clouds that had spawned and shaped the crystal were now largely spent. The amphitheater of fresh-fallen snow glowed faintly blue as the moon resumed its prominence in the night sky.

January sunrises arrive late, if at all, in the subarctic mountains. It was past eleven in the morning before the first glint of sunshine struck the surface of fresh-fallen snow. Around the column crystal were millions similar to itself—some stubbier, some longer—plus countless stellar crystals, hexagonal plates, needles, and irregular ice pellets. Each one had a unique history of creation and change during its fall from the sky.

As the sunlight grew stronger, this still blanket of mixed snow began to take on a wealth of winter colors. Having traveled unperturbed for 150 million kilometers (93 million miles), white rays from the sun now split into glittering pinpoints of sapphire, emerald, amethyst, and ruby. Before two o'clock the snow surface became flat and featureless once again, as the sun slid behind a knifelike ridge to the southwest. Stillness and shadow reigned over the alpine amphitheater.

Unaided by mechanical buffeting of the wind or thermal stress from the sun, the pristine crystals began to decompose. Mass self-destruction proceeded at various rates, depending on the frailty of each crystal. The finest stellar crystals were the first to go. In the exquisite architecture of their lacy arms lay the seeds of their own destruction. Their intricate frames and large surface area to volume ratio were energetically unstable, a condition which the laws of thermodynamics would not tolerate for long. Within 24 hours their filigree margins became rounded and smooth as molecules of water vapor moved from the tips of the arms and refroze toward the crystals' more stable centers. As if too beautiful to last, the stellar crystals contracted and collapsed. Over the next two weeks, their arms were reduced to even-edged blobs displaying all the elegance of bowling pins.

Meanwhile the column crystal, though structurally much sounder, also succumbed to nature's unending campaign against the orderly arrangement of molecules. The six well-defined edges of the column lost all their sharpness within a week of the crystal's landing. Within a month, its hollow core had collapsed in response to further energetic downshifts and the pressure of fresh snow layers above. By late January the former column crystal and all that fell with it sixty nights earlier had been transformed into old snow lying within the dense middle layers of the snowcover. Regardless of original shape or life history, all crystals were reduced, more or less, to the most energetically ideal shape available to nature: an undifferentiated sphere.

❄ ❄ ❄

Throughout the last half of February, the dry arctic air mass and warm, muggy Pacific air mass were locked in aerial combat as they struggled for supremacy over northwestern Canada. Their regular sallies back and forth over the mountains brought cold still weather one day, followed by blizzards the next. An even surface of fresh fluffy snow would drape the mountain only briefly at this time of year. Soon after falling it was blasted into a wind-crust, carved into dunes, or sublimated into vapor by cold thirsty air from the north. Or it was buried under 0.5 meters (1.6 feet) of heavy wet snow blown in from the south. The story of these weather fluctuations was recorded in the changing densities and textures of the snowcover's layers as truly as the rise and fall of oceans were recorded in the mountain's sedimentary rocks below.

Far removed from the wide flux of weather occurring at the surface was a round grain of ice that housed a microscopic fleck of volcanic ash from the Philippines. It had settled near the bottom of the snowcover, where heat flowing up from the earth warmed the air space around the grain to a steady $-5°C$ ($23°F$)—$20°C$ warmer than the prevailing temperature at the surface. The ice grain shrank in the heat as it gave up its water molecules to the warm, rising vapor. Finally it disappeared, releasing the ash particle to the surrounding snow.

Molecules of water vapor moved through the snow along a thermal gradient of ever-cooler temperatures. They vaporized, then froze again, passing upward from grain to grain in a hand-to-hand sort of fashion. New snow crystals were under construction, some of them growing quite large. These were pukak crystals that bore utterly no resemblance to any kind of fresh-fallen crystals or coarse old grains. Instead, they looked like hollow pyramids or pagodas built from minute logs of glass neatly scrolled together. In this way, bottom layers of old snow completely vanished and were reconstructed as pukak crystals, one of which had wrapped itself around the errant fleck of ash.

By mid-March there was real warmth in the noon sun as it bathed the mountain's jagged brow and most of the snowy theater below. The amplitude of daily swings in temperature increased dramatically. Nights of −30°C (−22°F) were followed by days of −5°C (23°F). Near the base of the snow-cover, thermal stability still reigned, though physical stability was fast deteriorating. The loose bonds between pukak crystals neared the breaking point as the heavier, denser layers above crept imperceptibly downslope under the incessant strain of gravity. This movement, now measurable in microns per hour, was building up forces which, if let loose, could exceed magnitudes of several tons per square meter.

Subtle shifts in surface tension caused spontaneous ruptures in the delicately poised pukak layer. Multiple fissures formed deep within the snowcover. Small random surface slides, moving no more than a few wheelbarrow's worth of snow at a time, gave expression to the dwindling mechanical strength below. Amid these stresses and strains, the pukak crystal now housing the ash fleck remained intact. It was, in fact, still growing new latticework though not as quickly as it had during the consistent cold of midwinter.

On the night of the March full moon, a translucent veil of scalelike altostratus clouds moved in over the mountains. By morning, the sky had turned a granite gray. The clouds thickened throughout the day and much of the next night until they finally opened just before dawn. At first they dropped cold dry snow in the form of fine needles and tiny bullet-shaped columns. Several hours into the storm, strong moist winds suddenly encroached from the southwest, changing the snowfall into pulpy wet pellets. In all, this one storm dumped 35 centimeters (14 inches) of new snow on the mountain in less than 24 hours. High winds and temperatures just below freezing continued long after the snow stopped, cementing the heavy upper layers into a thick cohesive slab. Though internally stabilized by the wind, the upper layers were now poised for mass motion on top of a hair-trigger platform of loose dry snow. Not far below was a stressed-out foundation of pukak.

The detonator was a half-ton chunk of limestone that popped off the mountaintop. One sunny afternoon in late March, a trickle of meltwater dribbled into a critical seam, joining boulder to bluff. That night it froze, muscling deeper into the crack as it expanded. The next morning, minutes after the sun struck the broken rock, the mountain let it drop. A union of 70 million years ended as the boulder gave in to gravity and made a brief rolling free fall through space. Still rolling, it careered down a snow-free talus slope of frost-shattered rock, bounced violently off another boulder, then disappeared with a soft thud into the mountain's snowy shoulder. It landed a few centimeters from the pukak crystal, which by now resembled a stretched and twisted staircase due to the mounting stresses from above. It was among the first of several billion crystals to shatter when the rock hit.

In rapid succession, dull booms rippled out from the impact crater as the pukak layer collapsed and let go of its weak hold on the underlying bedrock. All pent-up tensions in the huge snow slab above found release as it disengaged explosively from the mountain. Where it came unhinged, the snow slab was 200 meters (650 feet) across. It dropped en masse from the forty-degree slope, accelerating rapidly on a fresh dry bed of needles and bullets which offered all the resistance of ball bearings. At full speed, the avalanche rumbled down the mountain with the power of fifty diesel locomotives. It lost no momentum as it entered a fan-shaped chute just below the treeline.

Caught up in the wake of the avalanche, the smashed skeleton of the pukak crystal was carried 0.5 kilometers (0.3 mile) down the mountain. It finally came to rest on the flat valley floor buried inside a 5-meter (16-foot) -high mound of disintegrated snow. Twenty seconds passed before the echo of the avalanche's roar stopped bouncing off the surrounding mountain flanks. Within 2 hours the huge mound of snow at the base of the avalanche chute had fully recrystallized and turned as hard as rock.

❋ ❋ ❋

A series of late winter blizzards dumped 30 centimeters (12 inches) of snow on the mountain valley. The additional weight further compacted the mound of cementlike snow at the bottom of the avalanche chute. The snow mass became denser and more coherent as air spaces among the old rounded snow grains shrank and new molecular bonds formed between them. Deep inside the snow mass, the ash fleck was housed in a smoothly furrowed lump of clear ice the shape of a sheep's head. This irregular grain of ice was a far distant cousin to the six-sided column that originally transported the ash back to earth and the delicate pukak crystal that had carried it down the mountain. No longer a separate component of the snow, the lump had bonded with several neighboring grains. Together they displayed the shape of a cluster of grapes long past their prime.

Then, for the first time in seven months, the temperature at the snow surface rose above 0°C (32°F). On clear days, the sheltered valley floor, which had remained in shadow for half of that time, now received over 8 hours of direct sunlight. Although 80 percent of this light was reflected back into outer space, infrared radiation penetrated several centimeters into the snow, shifting the net radiation balance into the positive. The snow mass began to melt. Meltwater began percolating down through the snow during the day, then refreezing at night. The snow mass soon became saturated with water, becoming heavier by the day. The pores among the snow grains shrank further still, becoming almost microscopic. This process, continued over many months, would transform this footprint of an avalanche into a small but credible glacier.

But soon melting got the upper hand over freezing, and the glacier was not to be. Under starry skies, the temperature continued to dip below zero most nights, creating a crust of firm ice on the surface. During the first few hours of the morning, this clear window over the snow acted like greenhouse glass. The trapped heat below penetrated deeply into the snow, forming warm hollows bridged by an icy layer above.

By early afternoon, the ice crust vanished. Warm southwest breezes swept down the valley, causing widespread sublimation at the snow surface. As the downward flow of heat finally reached the underlying feathermoss, up came the snow fleas to engage in their erratic spring mating dance. Thousands of their black wiggling bodies peppered the surface, hastening the absorption of heat and the melting of snow. So did the dust particles, pollen spores, and other nuclei around which each original snow crystal had

Snowmelt en route to the Beaufort Sea.
Tessa Macintosh

formed. So did all the twigs, needles, and other windblown debris that landed on the surface. All these slowly sank as they absorbed heat, then radiated
it to the adjacent snow.

It was a lone pellet from a snowshoe hare that ultimately broke the temporary alliance between ash and ice. The grain of ice was now barely distinguishable from its neighbors, having all but merged with them during repeated floods of meltwater. Directly above it, heat radiating from the dark hare pellet had sublimed a vertical tube down through the snow. Just before the shadow of a mountain stole across the surface, the snow tube reached the level of the ice grain. Heat given off by the hare pellet was sufficient to mobilize the grain's constituent water molecules to a higher energetic state. Millions of molecular bonds broke within seconds as the ice grain vaporized into thin air. Ten days later, the ash particle it had once housed was flushed from the decaying mound of snow at the base of the avalanche chute. It came to rest briefly on a fire-charred spruce log that teetered over the bank of a small stream channel. Meltwater from south-facing slopes above soon flooded the winding channel, swamping the spruce log and carrying the ash particle into the headwaters of the South Nahanni River. Six weeks later it was halfway down the Mackenzie. On the last day of August it settled to the bottom of the Beaufort Sea.

Chapter 3
Living Snow

A patch of snow, in the wilderness or on a city street, is a laboratory for understanding ecology. Its lessons are simple. The living parts of our earth interact with the non-living. What happens to both in the finite time that the snow remains is a microcosm of larger ecosystem dynamics. An experiment in this lab is merely a winter walk with our senses alert to the changes taking place in ourselves and in the snow.

—Bernard Mergen, Snow in America[39]

Signals in the Snow

As much as I adore snow, I have little room in my heart for snowmobiles. It's not because they are boisterously loud, their infernal insect-like whines defiling, willy-nilly, several square kilometers of winter stillness wherever they go. It's not because when idling for more than 5 seconds, they create a pall of nauseating smoke as thick and impenetrable as the gas clouds over Jupiter. It's not because many drivers have a habit of brainlessly tossing beer cans, cigarette packs, and plastic oil bottles along the trail. No. None of these acts of desecration alone can explain my deep-seated disdain for snowmobiles.

What really gets me going, what causes my fists to clench and draws visible steam from my ears is the snowmobile's merciless and maniacal murdering of the living fabric of snow. You've heard of telemarking, landmarking, earmarking, and postmarking. Well, now there's highmarking. A friend of mine from the Yukon explained it to me this way: "A lot of crazy snowmobilers make a sport of going to the wildest, most pristine wilderness they can find, zooming up the highest mountain around, then carving it up with their tracks. The person who leaves the highest mark on the mountain wins. You wouldn't believe the risks they'll take to get high. You can see their marks for miles." Apparently they'll do this all day, answering the primal call of some Neanderthal obsession until somebody runs out of gas, breaks a leg, or gets swallowed by an avalanche.

Snow charioteer grabs extreme air. This might be one approach for minimizing the impact of snowmobiles on wildlife—at least until they touch down.
Northern News Service

I'm told that if there are no mountains around, the highest hill or haystack will do. In flatter country, the rules of the game change, the challenge now being to find the deepest, softest, machine-eating snow and then make your mark by thoroughly mutilating it without getting stuck. In my neck of the woods such snow lies along sheltered lakeshores—the very place where snow-adapted wildlife prefer to hang out.

Here caribou dig feeding craters through the soft snow to reach lush grasses and sedges below. Ptarmigan and grouse feed on abundant shoreline willows by day and tuck under an insulating blanket of snow by night. Weasels, otters, and mink use the shoreline as a highway between patches of ice-free water. And voles, shrews, lemmings, and mice, all busy beneath the snow throughout the winter, often reach their highest densities along lake or wetland edges—that is, until they are hammered from above by a snowmobile. Just *one* pass of a snowmobile can bring down the house on such ecologically vital species for an entire winter with a force of 400 kilograms per cubic meter (880 pounds per cubic yard), or roughly the same wallop as three sumo wrestlers dropping into a grand slam.

What motivates people to engage in such pernicious play eluded me until I read Edward Abbey's *Hayduke Lives!*

> The purpose of snowmobile recreation is not to get anywhere, see anybody or understand anything but to generate noise, poison the air, crush vegetation, destroy wildlife, waste energy, promote entropy and accelerate the unfolding of the second law of thermodynamics. For this purpose, then, an endless circling round and round from morn to night could be perfectly satisfactory to all participants. . . . Everyone knows that.[40]

Now, you may read this and decide that Edward Abbey is some sort of crackpot eco-extremist, a tribute for which he probably would have thanked you. But the next time you find yourself confronted by a roving horde of snowmobilers, particularly of the juvenile subspecies, take a moment or two to study their curious behavior. There is more than a smallish kernel of truth to what Abbey is saying. To me it's staggeringly obvious: today's shark-nosed, overpowered snowmobiles sow insanity in a majority of their drivers.

Having said that, I should tell you that I have been a loyal snowmobile owner for many years. Mine is not a very powerful machine. Nor is it very big. Many friends and neighbors have told me to my face that, parked on its go-cart-like body, my six-foot-four frame looks truly ridiculous. My snowmobile stinks and howls like all the rest. As for excitement, I get about as much tingle from driving it as I would from pushing a wheelbarrow full of horse manure. Still, as a simple beast of burden, it does come in handy now and then for hauling friends, fuel, and firewood out to my beloved backwoods cabin.

So there I was, a few weeks back, rattling along on my boxy little snowmobile. The snow was still at its warm and friendly fluffiest (as I now write, it is fast decomposing out my window). I had just delivered a sled-load of friends to their vehicle and was heading back to my cabin to paint Easter eggs with my kids. Out of habit, I stopped the machine every couple kilometers or so to realign my bones and tune in to the heady hush of the snow-clad forest.

Along a small lake protected by ramparts of tall spruce, I discovered an enticing stretch of shoreline untracked by any trace of snowmobile, snowshoe, or ski. The snow here was as fresh and immaculate as it gets. On its glittering surface I could read the prints of bird and rodent, hare and fox, but, as Edward Abbey would have said, "nothing man- or woman-made."[41] As I beheld the virginal mantle of white and light all around me, the first word that came to mind was *benevolent*.

> *Supple swatches of jewel-strewn snow conform to the grassy hummocks and boulders below. Rivulets of rainbow colors flash shy, sensuous smiles. I step off the snowmobile trail to admire some fox tracks splashing ninety degrees out from the shore. The regular, arrow-straight passage tells me that this fox knew exactly where it was going. I understand the magic in these tracks—in all tracks—not printed in the snow, not sketched, but sculpted. The deep, leaping, whole-body divots of this fox are true sculptures. They have three dimensions; they have texture and contour. They give form to motion. They exhibit different speeds. They reveal thoughts, decisions and moods. Frozen in these tracks is the mind of an animal immensely at home in the snow. Not merely "animal sign" or a record of behavior, there is something more here. Something magical.*

More signs and wonders awaited me at the far end of the lake, a water body my wife and I had come to know as "Bou Bog," for the caribou tracks that often pepper its surface. But I found no caribou tracks this day.

> *At the birch-lined end of the lake, I stop my machine again and bend low to peer into a thumb-sized tunnel immediately beside the trail. Down on all fours, nose to the snow, I play Peeping Tom on the pukak-lined world of the red-backed vole. The original tunnel has been sliced by*

the thundering treads of snowmobiles, forcing the voles to risk an open-air dash into the woods. While following its leaping four-print pattern over the snow, I hear the faintest hint of an engine, that hateful whining wail of a snowmobile. "How could anyone dare rupture this peace?" I thought, forgetting for a moment how I transported myself to this lovely spot. My first response is denial, and I continue to follow the vole's tracks—down this spruce bole, beneath that bent willow.

The drone rises. I have to finally admit that another snowmobile is coming right for me. It roars to a stop several meters from my silent machine, which blocks the trail. I lumber out of the woods, feeling at once embarrassed and haughty. Two riders on one machine. They just sit there. I marvel at how anonymous and positively arrogant anyone looks when wearing a snowmobile helmet.

Abbey calls them androids, "goggled, helmeted space-suited androids, encased within the screaming roar of their infantile machines, driving themselves onward, sealed off from everything but the red light, exhaust fumes and thrashing treads of the idiot in front of them."[42] But I digress.

I give the riders a limp wave, start my motor, and pull over onto some fresh snow, all the while imagining the collapse of a labyrinth of vole tunnels below. As they pass, shoulder to shoulder from me, I make out that they are genteel adults like me. Probably a harmless husband and wife team out for a Sunday joyride. So I lean closer and explain my situation.

"I was just looking at mouse tracks," I shout through my own Plexiglas visor. "They're everywhere!"

I intentionally say "mouse," assuming that the couple has no clue what a vole is. I feel it is important that these nice folks should know there are people in this world who get a real kick out of such idle pastimes as tracking small mammals in the snow. The riders nod mutely, smile, then putt down the trail. I make a point of looking back over my shoulder to see if they too are looking for "mouse" tracks. To my delight, both have lifted their visors and are peering at the snow beside the trail. Then they rev their machine and disappear in a cloud of shattered snow crystals.

A snowy sketch pad recording the free flow of animal traffic, in this case, a couple of barren-ground caribou.

Lorne Schollar

As I too roared away, in the opposite direction, I bathed in the smug satisfaction that I had converted a pair of snowmobile androids into believers in the secret life of snow. Vole-huggers, perhaps. On the final leg of my journey to the cabin, I steered my faithful snow chariot with extra care, sticking to the hard-packed trail, like a locomotive to rails, so as not to disturb any living thing.

Living Underfoot

Dark spruce trees burdened by snow belie the vibrant life that pulses below.
John Poirier

How badly do you want to skip outside and go birdwatching or tracking animals in the snow when you read grim winter sketches of the kind Jack London paints in his hoary classic, *White Fang?*

> Dark spruce forest frowned on either side of the frozen waterway. The trees had been stripped by a recent wind of their white covering . . . and they seemed to lean toward each other, black and ominous, in the fading light. A vast silence reigned over the land. The land itself was a desolation, lifeless, without movement, so lone and cold that the spirit of it was not even that of sadness. There was a hint in it of laughter . . . a laughter cold as the frost. . . . It was the masterful and incommunicable wisdom of eternity laughing at the futility of life and the effort of life. It was the Wild, the savage, frozen-hearted Northland Wild.[43]

Excuse me, Jack, but that's *my* backyard you're slamming. Thanks to the popularity of such forbidding portraits of winter, you might get blank looks

from most people when you suggest that snow is anything more than a sterile and suffocating shroud over the landscape. Few would argue that, for most
of the winter, life as we know it is clenched in a deathlike pose of suspended animation.

What about snow's redeeming values? At its twinkly best, who would
argue that it offers sweet candy for the eyes? How about a fetching substrate
on which to play our winter games? No quibbles there. But beyond whatever aesthetic or recreational values we humans may squeeze from it, what
good is snow to other life forms? Suggest that it is, in fact, a benevolent life-
supporting blanket on which many creatures depend for their very survival,
and your audience may quickly dwindle. The probability that any living
thing—from snow fleas to pygmy shrews—could actually be *down there,* alive
and kicking beneath the snow, may seem as likely as finding life on the
moon's backside.

Like so many ecological values (and this may spell our species' ultimate
demise), snow's inestimable worth to many North American plants and animals is essentially out of sight, out of mind. Winter puts on a good show for
us, masquerading as death while nurturing abundant life. Don't believe me?
Go browse through any of the many publications on snow by North
America's grandmaster of snow ecology, Bill Pruitt, professor emeritus at the
University of Manitoba in Winnipeg.

On campus they call him the Snowman or Dr. Snow or simply—but not
without respect—that snow guy. Even without knowing anything about him,
you might see Pruitt strolling easily through the halls of the zoology building
and conclude, from his appearance alone, that he hailed from some far northern land of snow and ice. The North Pole, perhaps. His stout form, gray beard,
ruddy complexion, and quizzical eyes rimmed with smile lines suggest forcibly
the image of an elf, if not Santa himself, disguised as a professor emeritus.

After devoting a half-century of winters to exploring the ecological
links between animals and snow, Pruitt knows as much about the subject as
any one person could. But this man is an island of knowledge surrounded
by a sea of what he calls "mass illiteracy and sanctioned prejudice regarding
snow." Quoting Voltaire, he complains that "here we have a country of
quelques arpents de neige—nothing but snow—but very little is known about
how animals and plants are affected by its characteristics."[44]

And so, he writes. Pruitt's intimate insights on snow blaze forth in his
popular prose. In his now classic *Wild Harmony: The Cycle of Life in the*

Northern Forest, Pruitt pioneered a unique writing style which places the reader behind the eyes, ears, and nose of that archetypical snow-dweller, *Microtus oeconomus,* colloquially known as the tundra vole. Jack London, take note. Even way out in northwestern Canada and Alaska where this creature lives, there's much more going on down there in the snow than meets the jaundiced eye:

> The new snow covered the layer of birch seeds and hid them from the small birds. The added weight of the fresh snow compacted the middle layers of the *api* [snowcover]. As the crystals squeezed together and broke, the cover creaked and groaned. A foraging vole would stop and huddle, ears twitching, and then resume its errand. The sounds breached the even tenor of life under the snow. An additional disturbance was the faint scent of birch carbohydrate that occasionally filtered down from above. Some voles dug upward through the layers of snow. One layer was easily tunneled, the next was harder; no two were alike. When a vole reached the seed-rich layer, it drove a horizontal drift along it and devoured every seed. Without the traditional scent boundaries to restrain them, some voles encountered their fellows directly. What squeaking and scuffling as the asocial animals repulsed invaders![45]

By tunneling his readers into the vole's cozy subnivean world, Pruitt reveals the critical dependence of such animals on snow. The farther north you go—in other words, the longer the winter—the truer this is. Snow is to the tundra vole what saltwater is to a shark. Without it the creature is dead. Writer Barry Lopez drives home this revolutionary notion in *Arctic Dreams,* a collection of thoughtful musings on the North:

> Winter, not summer, is more the season of record in the Arctic for the evolutionary biologist. A northern ecologist looking at snow sees an element as integral to the landscape as soil. . . . Snow [is] as fundamental in shaping an animal's life in the Arctic as rainfall is in the Philippines or sunlight in the Arabian Desert.[46]

While that may well be true for tundra voles in the frozen North, where winter calls the shots for over half the year, how important is snow to small mammals dwelling in more temperate regions, where winter is often no more than a seasonal bit player? Awfully important, discovered some of Pruitt's colleagues when they studied voles in the manicured forests of Europe.

Few Europeans who lived through it will ever forget the winter of 1962. From Scotland to Siberia, severe blizzards, deep snow, and record low temperatures severely crippled all forms of transportation for months. Schools were closed for weeks on end. Many senior citizens expired in their beds, succumbing to bone-numbing cold. Millions of people greeted Europe's snowiest winter in one hundred years with clenched teeth and raised fists. Meanwhile, unknown to all but a small circle of German zoologists, the region's vole populations, snugged in under all that snow, were enjoying unprecedented abundance and weight gains. The zoologists' data clearly showed that, where the snow lay deepest and longest, voles were breeding like rabbits and fattening like pigs.

Life is good down under, though often you'd never guess much was going on beneath the surface.
Lorne Schollar

Whether living on the shoulder of Alaska's Mount McKinley or in an engineered forest in Bavaria, small mammals dwelling under the snow depend on pukak, that warm, moist, loose layer of crystals that hugs the earth. There they find a thermally stable environment through which they can tunnel with ease. If the snowcover is deep enough, they may be virtually immune to predation by foxes, owls, or weasels. Thus ensconced all winter, a vole, lemming, mouse, or shrew can effortlessly maneuver through its maze of corridors, stopping at a food cache for a nibble of berries, checking its many sentry posts at the edge of its territory, or visiting a nest chamber to groom its fur and steal a quick nap. Never mind the bitter winds and cold snaps and predators that trouble surface-dwellers. For these furry little critters, life is good down under.

Bill Pruitt has a soft spot for pukak. "I am more fascinated by pukak than by any other kind of snowcover," he once told me. "You have to get down on your belly, necessarily with a magnifier and light reflector, in order to appreciate pukak. It is so fragile, so beautiful and so important in the lives of small herbivores, small carnivores and invertebrates." Long ago Pruitt concluded that if it were not for the pukak layer, large parts of our northern forests would be devoid of small mammals. They are physiologically incapable of surviving winter anywhere else. The ecological implications of this scenario are unthinkable since so many food chains are propped up somewhere along the line by small mammals, whose lives, in turn, are sustained by a beneficent blanket of snow.

Flea Circus

Near the back of a dusty 1885 issue of *Harper's* magazine, you can read about American botanist William Gibson's cosmological insights, apparently inspired by minuscule creatures that creep and crawl over the snow. In his testimonial "A Winter Walk," he encourages you to "lie down upon the snow and shut out the distant trees, divest [yourself of your] physical identity, and look up at this beetling range as an ant might do. . . . At this focal range . . . man learns his true status as a constituent of the universe."[47]

Gibson may have been just pulling our metaphysical legs here, but I happen to take his advice quite seriously. Adopt a prone and humble stance on the snow, preferably faceup, and indeed you will discover that you are not alone down there. I'm not talking about obtrusive police officers, though that is always a hazard for committed snow inspectors. Anyone publicly preoccupied with a substance as banal as snow risks suspicion, if not charges of outright flakiness. I'm talking now about snow fleas, snow scorpions, ice worms, that sort of thing.

Where better to start than with snow fleas? From Salem, Massachusetts, a place infamous for burning witches and kindling superstitions, comes the newspaper headline "Warm Winter Brings Fleas." The story tells of a woman who became alarmed when the snow in her front yard spontaneously erupted in huge patches of black specks. "When she looked more closely," wrote a brave reporter, "she got the willies—the specks were hopping and jumping." She took one inside to examine under a magnifying glass and discovered what the reporter called "one of the more unusual torments of our winters—a snow flea."

"I didn't know who to call," she said. "Let's face it. What would someone think of a woman complaining of billions of fleas on her snow?" In the end she consulted a professor friend of hers who assured her that "they're not doing any harm, but they can be kind of annoying. The cold won't hurt them," he added, in kindness to the fleas. "They'll probably just go back under the snow."[48]

Behold the Arthropod "Iditarod" International Sled Race. The sleds are powered by several kinds of snow fleas and skippered by a snow scorpionfly (foreground), winter snow fly (airborne), and Leiodid beetle (between). In the background is a sled pulled by unruly dog fleas that have sent their master—a winter tick—flying headlong into the snow. Snow-loving predatory mites heckle from the sidelines. The shining face of a winter stone fly looms in the sky.
Barry Flahey

Now, of course if you're no bigger than a fragment of crushed peppercorn and spend nine-tenths of the winter hidden beneath the snow, you're bound to be misunderstood, especially in a crucible of controversy like Salem. Stir in our customary aversion toward anything to do with fleas (hence *fleabag, fleapit, flea-bitten),* and the odds of a fair and impartial assessment of your worth are rather low. Now sprinkle together the words *insect* and *winter,* and you've got a knee-jerk recipe for rancor and revulsion.

Much more than a pestiferous pipsqueak, the meek and mild-mannered snow flea is neither a flea nor a true insect but a springtail. And unlike mosquitoes and blackflies, the snow flea has not the slightest interest in human flesh and blood. In truth, snow fleas pose about the same threat to us as earthworms. Most are strict vegetarians, eating microscopic fungi, algae, and plant material that passively fall on the snow or blow in from the nearest mountaintop.

In Washington state they call it the glacier flea after its habit of blackening the icy flanks of Mount Rainier. Its European cousin goes by the same name, only there you say *gletscherfloh.* Take your pick. One mountaineer reported that by day "they live off nothing but conifer pollen blown onto the glacier. At night they freeze fast to the firn and ice." Their abundance at high altitudes is frankly unbelievable. Imagine living out your whole life on permanent snow and ice on a 6000-meter (20,000-foot) -high Himalayan peak. That's where mountain-climbing entomologists are finding them. Here snow fleas subsist on heavenly manna: windborne pollen grains, microorganisms, and other organic fallout. There is some speculation that the more junk we stir up in the air through deforestation and agriculture, the higher snow fleas will climb, or rather, jump.

More kindred to lobsters than blackflies, snow fleas and other springtails are members of a prestigious arthropod order, the Collembola, known mostly by their wingless, multisegmented bodies and pogo-stick tails. (Hence Mount Collembola in Alberta's Eastern Slopes; they say it virtually hops with fleas.) Latched like a mousetrap under the snow flea's belly is a long forked plate, a furcula, which, when released, thwacks down hard against the supporting surface, pitching the snow flea across distances hundreds of times its own pencil-point length.

My erudite textbook on snow ecology, entitled (you guessed it) *Snow Ecology,* reports that "in fair weather," jumping resolutely in one direction, a snow flea can fling itself forward almost 300 meters (1,000 feet) per day or

more than 3 kilometers (2 miles) in ten. If you could jump to the same proportions, vaulting over a twelve-storey building in a single bound would be
a piece of cake. In ten days of fair-weather hopping, you could easily take in
a New York opera and the San Diego Zoo in the same, quite affordable,
vacation.[49]

In his *Handbook of the Canadian Rockies,* mountain man Ben Gadd
swears he's heard a party of snow fleas whooping it up in flight. "Listen
closely to an airborne Collembola; you may hear it saying 'wheeeeee!' "[50]
Now, as much as you will thoroughly enjoy Gadd's book, keep in mind that
this man spends a lot of time (perhaps too much time) in thin air.

If you've never had the good fortune to see them, let alone hear them,
I couldn't blame you for lumping snow fleas in with unicorns, centaurs, and
other creatures that dwell only in the imagination. In fact, their biological
credentials are as sound as those of seahorses and winged squirrels, but
finding snow fleas is much easier. That is, if you know where and when to
look for them. As Salem residents discovered, climbing ropes and ice axes are
not necessarily required.

Who better to turn to for advice than Asa Fitch, the snow flea
aficionado who first put these critters on the scientific map? Working as
official entomologist for the state of New York, he left us these vivid field
notes from the winter of 1847:

> This is an abundant species in our forests in the winter and fore
> part of spring. At any time in the winter, whenever a few days
> of mild weather occur, the surface of the snow, often over whole
> acres of woodland, may be found sprinkled more or less thick-
> ly with these minute fleas, looking at first sight, as though gun-
> powder had been there scattered. Hollows and holes in the
> snow are often black with multitudes. The fine meal-like pow-
> der, with which their bodies are coated, enables them to float
> buoyantly upon the surface of water, without becoming wet.
> When the snow is melting so as to produce small rivulets cours-
> ing along the tracks of the lumberman's sleigh, these snow fleas
> are often observed, floating passively in its current, in such
> numbers as to form continuous strings; whilst the eddies and
> still pools gather them in such myriads as to wholly hide the
> element beneath them.[51]

Then there's Frank Bolles, who, wandering the flea-bitten hinterlands of Harvard, closely inspected and described the snowy tracks of everything from feral cats to his neighbor's toy sled. Imagine the boyish glee of so curious a man when he blundered upon swarm after swarm of mating snow fleas. He wrote about this "discovery" in his 1891 book *Land of the Lingering Snow.*

> While recrossing pasture and field, swamp and thicket, I noticed countless black specks upon the snow. They moved. They were alive. Wherever a footprint, a sharp edge of drift, or a stone wall broke the monotony of the snow surface, these black specks accumulated, and heaped themselves against the barrier. For miles every inch of snow had from one to a dozen of these specks upon it. What were they? Snowfleas or springtails *(Achorutes nivicola),* one of the mysteries of winter, one of the extravagances of animal life.[52]

In late winter, whether out in the woods or in your rose garden, when the sun is high and the thermometer hovers around 0°C (32°F), you too will find the dark multitudes appearing as if out of thin air. Or try a warm sheltered hollow near the base of a rock or tree. Check yesterday's bootprints or ski tracks. You'll see snow fleas boinging over the snow. In the mountains you may detect their rainbow colors: yellow, red, orange, purple, and blue, as well as brownish black. But wherever you spot them, keep a respectful distance. As Bolles discovered, this is mating time for snow fleas. A virtual arthropod orgy.

Whatever hopping, wooing, and strutting goes on down there on the snow, it never culminates in embrace. The males and females may not even see each other, yet fertilization takes place. At some magic moment, the male secretes a packet of sperm on a tiny stalk shaped like a golf tee. When the female comes along, she scoops up the sperm packet and inserts it into her private parts. A few weeks after the eggs are laid, they hatch on warm snow-free soil. The newly emerged nymphs immediately get down to the business of soil building which, for most snow fleas, is their true calling in life.

Seen or unseen, we really owe a lot to snow fleas and their springtail kith and kin. Without them, our soils would be much impoverished, particularly in arctic and alpine settings. These lowly creatures are pivotal players in recycling plant nutrients and creating soil. By the billions, they grind up dead organic matter into minute particles that only then can be attacked by fungi and bacteria, which carry these vital processes to completion. In mountainous areas, where snow fleas spend most of their days on ice, they are devoured by a host of mites, spiders, and other predatory creatures lurking on the snow, which in turn are eaten by larger beetles and bugs, which in turn feed small mammals and alpine songbirds, which in turn feed ravens and high-flying hawks . . . You get the picture. Snow fleas are important.

Any sign of new life near winter's end will gladden the heart of even the most hard-nosed snow hugger. Looking for mating swarms of snow fleas ranks high on my list of vernal rituals. Discovering them assures me that our soils are in good health and that winter, as much as I love it, is about to make way for spring. Hence my belief that the emergence of snow fleas deserves much wider recognition. Up in the subarctic, where I live, when the snow begins to melt, we have no groundhogs out looking for their shadows. Nothing against groundhogs, but I believe a more inclusive kind of celebration is in order, one that snow inspectors from Tuktoyaktuk to Tennessee can observe in concert with the retreat of their particular winters. And so, may I propose we join hands and institute Snow Flea Day? I'll get back to you on this.

Indigo Worms and Watermelon Snow

Yes, Virginia, that *is* an ice worm crawling over your boot. Contrary to popular opinion, this unassuming denizen of snow and ice is no more fictitious than a snow flea. The ice worm owes much of its claim to fame to that gold rush bard of the Yukon, Robert Service, who immortalized this creature in his "Ballad of the Ice-Worm Cocktail."[53] In the same chilly breath, he managed to cast a thick veil of doubt over its very existence.

Amidst the smoke and din of Dawson City's Malamute Saloon in the far-flung Yukon, we meet the gallant Major Brown, whose habit of "sticking out his chest as if he owned the town" prompts the local Arctic Brotherhood to challenge this newcomer to the Ice-Worm Cocktail Test. Leading the challenge is local ice worm authority, Deacon White, who, in answer to Major Brown's understandable queries, provides some insightful details on the ice worm's habitat preferences:

> *"Yet (pray excuse my ignorance of matters such as these)*
> *A cocktail I can understand—but what's an ice-worm please?"*
> *Said Deacon White: "It is not strange that you should fail to know,*

Since ice-worms are peculiar to the Mountain of Blue Snow.
Within the Polar rim it rears, a solitary peak,
And in the smoke of early Spring (a spectacle unique)
Like flames it leaps upon the sight and thrills you through and through,
For though its cone is piercing white, its base is blazing blue.
Yet all is clear as you draw near—for coyly peering out
Are hosts and hosts of tiny worms, each indigo of snout.

We can only conclude that White was a keen observer of ice worms, for he goes on to describe their extraordinary foraging strategies and subtle movement patterns on the inhospitable flanks of this lone blue mountain.

"And as no nourishment they find, to keep themselves alive
They masticate each other's tails, till just the Tough survive.
Yet on this stern and Spartan fare so rapidly they grow,
That some attain six inches by the melting of the snow."

After a suspenseful moment when Barman Bill declares that he's fresh out of pickled ice worms, he remembers a secret stash in a drawer behind the bar. Here, the specimens that he unveils speak volumes about the ice worm's unique anatomy.

"Yet wait . . . By gosh! It seems to me that some of extra size
Were picked and put away to show the scientific guys."
Then deeply in a drawer he sought, and there he found a jar,
The which with due and proper pride he put upon the bar;
And in it, wreathed in queasy rings, or rolled into a ball,
A score of grey and greasy things were drowned in alcohol.
Their bellies were a bilious blue, their eyes a bulbous red;
Their backs were grey, and gross were they, and hideous of head.
And when with gusto and a fork the barman speared one out,
It must have gone four inches from its tail-tip to its snout.
Cried Deacon White with deep delight: "Say, isn't that a beaut?"
"I think it is," sniffed Major Brown, "a most disgustin' brute."

At the point where Major Brown, having watched "with horror" as his challengers dashed their cocktails down, reaches for his own glass, we gather some final details on the behavior, waste products, and countenance of this inscrutable creature.

The Major gripped his gleaming glass and laid it to his lips,
And as despairfully he took some nauseated sips,
From out of its coil of crapulence the ice-worm raised its head;
Its muzzle was a murky blue, its eyes a ruby red.

You can understand how a poem like this might spread a little confusion about the ice worm's very existence. Certainly Major Brown had his doubts. This matter was partially put to bed when somebody actually pounced on one and stuffed it into a bottle of formaldehyde for all the world to see.

During the summer of 1913, University of Toronto biologist Edmond Walker was on a field trip to Banff, Alberta, to check out its enticing hot springs. His notes give no hint as to whether this was a vacation masquerading as a scientific expedition or vice versa (ask me about my trip to Hawaii sometime). What jumps off the pages of his field book, however, is the discovery of what he called "icebugs" in the late spring snow just above the luxurious Banff Springs Hotel (his presumed field camp).

As curator for one of the world's most prestigious invertebrate collections at the Royal Ontario Museum, Walker felt obligated to upgrade the name of this controversial creature. "Icebug" just would not cut the mustard among his peers. Unsure whether its genealogical roots sprang from the family of crickets—Gryllidae—or cockroaches—Blattidae—he hit upon Grylloblattodea, the label for an entirely new order of insects hitherto unknown to science. Much to Walker's delight, this humblest of bugs was later chosen as the official emblem for the Entomological Society of Canada. "It's a perfect choice," one entomologist told me. "It's a very Canadian insect since it lives in the snow and is deathly allergic to heat."

Skeptics, take note. Though nowhere near the spaghetti-sized specimens in Robert Service's poem, ice worms, in fact, come in about a dozen colorful models, from yellowish brown to red, all distinctly wormlike. The search goes on for the "bilious blue" species documented by Robert Service. The dark brown specimen collected by Walker was actually the larva of a flightless insect known on the street as a rock crawler. The other more common ice worms are lumped in the genus *Mesenchytraeus*. These are true worms, cousins of the ones in your garden. Pinned down to a laboratory tray, they are about the length of your thumbnail. But usually you'll find them curled up into little balls (reminiscent, perhaps, of Robert Service's "queasy rings"), which is not at all surprising given where they live. Would you not do the same?

As famous as they are, not much is known about how ice worms get by. Two quite reputable sources paint vastly different portraits of their life cycle. In the bug section of his *Handbook of the Canadian Rockies,* Ben Gadd says that they "seem to live only on snowbanks that melt entirely each summer, and only on snowbanks lying over soil (not snow lying over glacial ice or rock). Eggs laid in soil hatch in early spring."[54] In this scenario the newly hatched wormlings are thought to work their way up to the snow surface, where they hang out till the late summer melt returns them to the soil.

In an altogether different scenario, glaciologist James Dyson states in *The World of Ice* that ice worms shun the soil entirely, spending their full life cycle on snow or ice. "Their eggs, deposited in snow, ice, or pools of water on the ice, must hatch in freezing temperatures," he concludes.[55] What *is* clear to me is that the confusion about these creatures persists—not undeservedly so.

However dimly science may understand the secret life of ice worms, we do know that ice worms are usually found on the high western slopes of mountains from Oregon to Alaska. I found nothing about their presence in the perpetually frozen mountains that rim Canada's Arctic. If you're ever poking around up there and trip over some ice worms, do let me know. One thing's for sure: if any creature could live up there, it would be the ice worm.

It turns out that an ice worm's favorite food is watermelon snow, yet another misunderstood life form that calls snow home.

Deep in the Coconino National Forest of Arizona is a place that owes its name to the pink-painted snows that festoon the pine woods every spring and summer: Raspberry Springs. Ski or snowshoe through this dreamlike snowscape, then look back at your tracks. Pink turns to red when compacted by your passage. This effect becomes magnified in the heat of the alpine summer, when you'd swear the meltwater rivulets are flowing with burgundy wine. Taste it and you may detect a faint hint of watermelon—at least so some bush people claim.

What is behind all this ruby-rouge snow? A blue-green algae dubbed *Chlamydomonas nivalis.* If, like me, you enjoy dissecting scientific names, I can tell you that *nival* means "of the snow." As for the other part, all I can make out is that it roughly translates into "clammy deity," but I may be grasping at straws here. But what's this? A *red* blue-green algae? Yes, this species does contain green chlorophyll, the same medium found in plants that lets them create food from mere sunlight. This algae's distinctive color comes from the red pigment in a gel-like casing that protects each microscopic cell from the fierce ultraviolet rays bouncing around the snow surface in all directions. Biologists like to use words such as *ancient* and *primitive* when describing this life form, believing that it had what it took to tough it out when, more than three billion years ago, our planet virtually hummed with levels of solar radiation under which you or I would sorely fry.

As excited as you may be to discover watermelon snow, life is not what you could call electrifying for little *Chlamydomonas*. What gets them trucking to the surface is an inviting flood of spring sunlight that penetrates the rotting snow. According to snow algae guru Ronald Hoham, a biologist from New York's Colgate University, this likely triggers the germination of snow algae "resting spores" at the soil–snow–interface. They sprout tails—"locomotory flagella"—that they use to beat their way up, up through the mini-cascades of meltwater to the light. After all that, they have little to look forward to in the food department. Besides chewing on raw sunlight, they get other essential nutrients from dust and motes of rock that lightly season the snow. Hoham believes that the depletion of such nutrients is actually "to the advantage" of these organisms, since it may in turn trigger their sexual processes. Put less prosaically, this means that the red-stained hordes, starved of their daily multiple vitamins, sublimate their hunger into carnal knowledge of one another. Hoham admits that such complex connections are still "under investigation." Whatever turns them on, the result is a well-behaved brood of cold-hardy resting spores.

Some years, red algae simply refuse to show up. It's not that they are fussy eaters—hardly. It has more to do with needing just the right combination of moisture, light, and temperature for their brief fling on the surface, which, on a good year, may last for a couple weeks during the peak melt. Hoham believes that "the resting spores become the survival stage on soil or over the forest floor once the snow disappears. These spores remain dormant for about fifty weeks until the cycle is repeated the following year."[56] Did you get that? That's fifty *weeks* of dormancy. Like I said, not a lively bunch, but tough.

In good years, when conditions are favorable and watermelon snow adds fleeting swatches of pink to the high forests and mountains, another bonus may await you. Those same bright warm conditions favor the growth of more familiar life forms. As a result, one day you may find yourself frozen in your tracks by a blushing patch of late-lying snow fringed with lush grasses and alpine wildflowers in full bloom.

But what's this *green* snow all about? And orange and yellow? Yes, it seems snow algae, too, come in an array of mind-bending colors. Green snow is caused by the red alga's self-effacing cousins called *Chloromonas,* literally meaning "green thing." Rather than creating an ostentatious display out in the open, these algae prefer to duck under a spruce or fir tree, or tuck into deeper snow layers, where solar radiation is less irksome. These algae often form dense emerald-colored bands caused by gradual compaction of the snow over the winter. This process can create a startling effect in springtime when these bands coalesce as meltwater begins percolating downward, causing the snow to glow green with concentrations of algae reaching a million cells per milliliter. Looking at such concentrations from above, you'd swear there was grass down there trying to burst through the snow.

Among the *Chloromonas* clan are orange and yellow algae that contain the same pigment that tints carrots and pumpkins. They are actually quite common, found in the mountains all the way from northern Mexico to Alaska. What's weird about them, though—what snow historian Bernard Mergen calls "perhaps the strangest ecological discovery of recent times"[57]— is one species' exclusive affection for ski hills.

Back in the spring of 1992 and '93, Ronald Hoham and his microbe mates were out a-sampling the snows of New England, when what should they find but two new species of *Chloromonas.* This in itself was cause for celebration, but imagine their surprise when, back at the lab, they realized that one of these, designated *Chloromonas* sp.-B, had been collected nowhere else but the highly engineered snows of such ski Meccas as Vermont's Mount Killington, Jay Peak, Mount Mansfield, and Stowe Mountain. After much speculation and, I suspect, several beers, Hoham figured that this species may travel from ski slope to ski slope on the bottom of skis, surviving stuffy closets and dank basements between winters. Ask Hoham about his pet name for this new species, and he'll tell you it's *Chloromonas skiensis.* I've yet to find this in the scientific literature, but maybe I'm not digging far enough. As for any yellow snow you might find out there, I suggest you give the likes of Ronald Hoham a call before getting your nose too deep in it.

Sex in the snow is apparently no problem for these versatile creatures, but the question of exactly how they do it had long puzzled Hoham. So, about a dozen years ago, he and four of his graduate students rounded up an eclectic bunch of snow algae from the mountains of Quebec, Arizona, and New York, introduced them to each other in a sterilized walk-in mating chamber, fed them on prechilled growth medium, laid them out on white enamel shelves, set the lights just right, and then watched. Out of 92 thousand rigorous observations, they managed to catch *Chloromonas* in the act a grand total of nineteen times. That's a little less than one score in five thousand peeps.

What they witnessed was downright shocking. In a normal mating procedure, sexually active algal cells will bump their heads together or, if feeling especially romantic, entwine their flagellating tails to culminate in a derriere embrace. While under the laboratory spotlight, however, these algae displayed an altogether different suite of positions. Among their deviant sexual behaviors Hoham cited the following: "V-shaped formations with posterior fusions, lateral–lateral fusions with anterior ends pointed in the same direction or in opposite directions, posterior–lateral fusions, and in rare cases, posterior–posterior fusions." And, if you can handle this, there is more. "In our studies, triple fusions were observed in very rare cases, and this involved the fusion of three sex cells at the same time." Such aberrant mating practices invariably resulted in aborted cells. The last resort in such cases is asexual reproduction, which not only reduces conviviality among snow algae but also their genetic diversity. This led Hoham to suspect that if they insist on carrying on like this, "some populations of snow algae may be headed toward evolutionary extinction."[58]

Maybe it was those enamel shelves that threw them off established sexual norms. Or the lights were all wrong. But in the field, at least, snow algae seem to be doing just fine.

Whatever their color or sexual preferences, it may be that, as with snow fleas, we owe a lot to snow algae—an *awful* lot. Just think about where and how they live. These are among the hardiest creatures on our planet. I told you about their dietary needs: sunlight and dust. Beyond that, all they need is a few drops of water for migration and mating. In hard times, they can slip into indefinite dormancy. On the other hand, in good times a handful of snow algae can reproduce like Yukon sourdough, eventually covering a hillside.

But most of the time these microbes live out their quiet lives in some of the most severe conditions doled out by nature: prolonged frigid temperatures, intense solar radiation, high acidity, minimal nutrients, drought. And there's more. Studies of snow algae from the mountains of Washington to the icecaps of Greenland indicate that they unwittingly concentrate high levels of heavy metals and other pollutants released by our own overindulgent species— another good reason not to eat colored snow. One of Hoham's colleagues honestly believes that dormant snow algae could be lurking at the bottom of the miles-thick Antarctic ice sheet. Like I say, these are tough little beasts.

The vicissitudes of such extreme habitats are not unlike what you would find in outer space. Is it any wonder, then, that NASA scientists have taken a keen interest in snow algae? Is it possible, they ask, that such microbes represent some kind of living analogs of ancient Martian life that, long ago, survived an interplanetary asteroid ride to a sterile Earth only to colonize the planet soon after impact? Could it be that *all life as we know it* owes its very existence to the great-great-great-etc. ancestors of today's snow algae? Something to think about, anyway, the next time you find yourself pondering a patch of yellow snow.

Ecologically speaking, the buck stops with snow algae. Think of them as the pasture grass of a food pyramid built on snow. They offer, as one writer put it, "an ice cream parlour of flavours in the food chain." These are the snowfield's primary producers, grazed upon by protozoa, rotifers, snow fleas, earwigs, and other bovine invertebrates.

Take the snow cranefly, for instance. Next time you are schussing through the subalpine, you may see it plodding slowly across the snow. You might mistake this dark wingless bug for a spider, but look again: only six legs, not eight. Like everybody else upon the snow surface—excluding more reserved skiers—snow craneflies spend much of their time mating, which for the males means certain death, and for the females, a final bout beneath the snow to lay their eggs in the soil. Exactly how snow craneflies resist freezing solid is a bit of a mystery. At temperatures below −15°C (5°F), they slip into what entomologists call a "chill coma." But thanks to a kind of antifreeze in their bodily fluids, they're still not frozen. Trouble is, in this supercooled state, they are subject to instant and complete freeze-up if jolted, which may explain why they seem to walk with such infinite care.

Many of these creatures actually *thrive* on cold temperatures. Remember the rock crawler, the squirming larva that got the ice worm folks all excited? Though now famous, little is known about its adult form other than that it is nocturnal and cold-seeking, unable to tolerate temperatures above 8°C (38°F). Place a rock crawler in your warm hand or breathe on it, and it croaks. Then there's the carabid snow beetle found in the highest mountains of Norway. Researchers kept the poor thing encased in solid ice for six months (starved of oxygen, too, I might add), then watched it merrily amble away when released from its icy tomb. You have to wonder if such creatures also might have hitched a ride from Mars a while back.

Upper-Crust Chionophiles

You won't find it on any map: a thin white line beyond which snow becomes a critical factor in determining the fate of plants and animals, big and small. My *Snow Ecology* textbook tells me that on any given piece of land, be it a mountain peak or desert basin, if 15–25 centimeters (6-10 inches) of snowcover persists for at least two to eight weeks of the year, snow rules. Beyond this magic threshold, animals must adapt to the demands imposed by snow, or die.

When you think about it, snow really can be a bother. For many animals it smothers food, impedes movement, and increases risk of predation. The same substance that can protect plants from killing frosts and drying winds also can wreak havoc on their world. Heavy snow can seriously maim if not kill trees. Late-lying snow can abort reproduction in flowering plants. And, when blasted by the wind, snow shrapnel can atomize any shoots or buds that dare expose themselves to the winter air.

All these mortal hazards can be exacerbated by a good snowstorm. This effect was amply shown by a classic study done by Stanford University ecologist Paul Ehrlich. In late January 1969, a hair-raising snowstorm conveniently struck near the Colorado field station where Ehrlich had been doing some snow ecology studies. Once they could see out the laboratory window, he and his colleagues rushed out to begin measuring the effects of the storm on the ecosystem's key plants, insects, and mammals. They discovered that the impact of this one storm was extensive and long-lasting. Flowering plants and grasses essential to many of the resident mammals were seriously damaged by excessive, prolonged snowloads. That year, almost 100 percent of the purple lupines they studied produced nary a seed; 80 percent of the corn lilies never even flowered. This damage sent negative ripples far and wide through insect and mammal food chains, several populations of which declined drastically. It was an outright disaster for one species of butterfly, which eventually became extinct within Ehrlich's study area.

When the snow begins to fly, clearly some animals are readier for it than others. Why is snow welcomed by some and deplored by others? This was a question pondered fifty years ago by Russian biologist Alexander Formozov. He spent many long hours watching animals engaged in a life and death struggle with snow. From the snow fields of Siberia to the frosted slopes of the Ural Mountains, he observed animals in the wild and concluded that, for some, snow poses one of the most profound environmental challenges they face. "Snow cover, for many species, is the most important element of environmental resistance," wrote Formozov, "and the struggle against this particular element is almost beyond some species' ability."[59]

With snow-white feathers, snowshoe feet, and a habit of diving under an insulating blanket of snow on cold nights, these willow ptarmigan are consummate chionophiles.

John Poirier

He carried this notion further by grouping animals according to their ability to cope with the demands imposed on them by snow. He recognized a spectrum of categories based on how well adapted to snow an animal may be. At one end, the maladapted end, was the chionophobe—from the Greek *chion* (KEE-on), "snow" and *phobos,* "fear"—literally, snow haters. (Can't handle a little wind? You're a pneumaphobe. Allergic to loud music? Audiophobe. You get the idea.) In this category he lumped small ungulates like his native steppe antelope, many sparrow-sized birds, and small cats.

Avoidance is how these animals cope with snow; if that fails all that's left to them is outright denial. Watch a housecat shake that hateful snow off its little paws, and you'll know what a chionophobe is.

At the other end of Formozov's snow spectrum are the chionophiles—*philos* being Greek for "dear" or "loving." The snow lovers. These animals are so well adapted to snow that their very lives may depend on it. Every species Formozov put in this category, which he affectionately called Chionees, lives in regions of "hard winters with much snow."[60] These include such northern icons as the ptarmigan, snowshoe hare, arctic fox, and collared lemming, whose front paws sprout into little snow shovels each winter.

Formozov built a number of gradations into his snow-coping yardstick, putting mice, wolves, moose, and wolverines, for instance, somewhere near the middle. These he called chioneuphores—from the Greek *phero,* "to bear"—animals whose tolerance to snow allows them to take it or leave it as they please.

As much as biologists like to impose order on nature (and I am no exception), the other day I heard an amazing story about a wolverine that made me think that, if Formozov were still around, I would recommend he upgrade that species to full chionophile status. The story came to me more or less over my back fence. (There's something about my earthy neighborhood that seems to attract science types. Next door is a hydrologist who is married to a permafrost expert. A polar bear biologist lived across the street from me until replaced recently by an arctic anthropologist. It was the grizzly bear biologist down the street who told me the wolverine story.)

Out in the barren-lands of Canada's Northwest Territories, John Lee had been tracking a wolverine all morning with little success. His mission was to tranquilize the animal and slap a radio collar on it, to help understand how these elusive creatures survive the arctic winter. But the wind-hardened snow gave him no clues to follow, being too firm to create much of a trail behind a retreating wolverine.

"That stuff is like concrete," Lee told me. "You can jump on it with all your might and not even make a dent." According to Formozov, wind-packed snow on the tundra is so hard that an axe when knocked against it rings as if it were struck against iron. He recommended that the compacted snow of the drifts be cut with a hand saw, "because the iron shovel is of little use."

Bumping along on a snowmobile (and I do mean bumping), Lee finally picked up a decent wolverine trail, displaying its distinctive five-toed tracks with wide arching palms. In a few sheltered spots, the snow was soft and deep enough for him to make out the distinct belly-drag pattern typical of a loping wolverine. But most of the snow was rock-hard.

Compacted wolf tracks stand out in bold relief against wind-sorted snow. Classed as chioneuphores, wolves can take or leave snow as they please.
Lorne Schollar

"After all that tracking, I almost freaked out when I came over a hill only to see the wolverine jump on the snow a couple times, then disappear. Somehow it dove right in!"

Lee spent the next few minutes trying to track the animal, not with his eyes but his ears, much as a predatory owl would listen for voles and mice skittering beneath the snow. "It was very weird. You could hear him scratching around down there like a giant mouse." The wolverine claimed victory after Lee's many heated attempts to dig him out and poke him with a tranquilizer (safely attached to a long pole). "At one point he took a chomp on my shovel. Then he exploded out of the snow and took off like a racehorse."

What most impressed Lee about this whole escapade was how the wolverine knew exactly where, in a land dominated by rock-hard snow, to gain access to the diggable layers below. "I went back to examine the spot where the wolverine had plunged in and discovered that it had taken advantage of the loose crystals around the stiff grasses sticking out of the snow."

I later did a little digging myself, and from a technical report entitled "Snow in the Biotic Environment," I learned that where plant stalks penetrate the snowcover, they act as thermal conductors, transferring heat from the warm earth to the atmosphere. On its way up, this heat creates a wide areola of loose depth hoar or pukak crystals around the stem, even in the hardest of snow. As Lee pointed out, "That wolverine obviously had a very precise knowledge of tundra snow conditions."

Animals adapt to snow in three ways: through their bodies, through their behavior, and through their brains. Judging by the wolverine's lack of any onboard accessories like snowshoe-sized feet or a snow-white coat, I would say that it was well endowed in the brains department. The snowshoe hare, in contrast, owes a lot to its remarkable body.

Foremost among Formozov's criteria for branding a species as snow lover or hater is its ability to move through snow while searching for food, escaping from predators, or generally just getting around. Sometimes deep and fluffy, sometimes hard and crusty, sometimes wet and mushy, snow, in its many forms and phases, presents a vast range of possibilities that may either help or hobble an animal engaged in the simple act of locomotion. In this realm the snowshoe hare, *Lepus americanus,* is the undisputed king.

This icon of winter crouches silent and still on the evolutionary summit of nature's adaptiveness to snow. The snowshoe hare solves the problem of locomotion in a snow-dominated landscape with its enormous hind paws which, as its common name suggests, act exactly like snowshoes. They are covered in long stiff hairs that almost double the size of the flesh-and-bone foot within. This design permits nimble passage over the softest, deepest snow in which the hare's predators often flounder. More than providing traction, the hare's marvelous feet provide impressive speed. Down a semipacked trail, a snowshoe hare can vault along at over 40 kilometers (25 miles) per hour, covering as much as 3 meters (10 feet) in a single bound.

The snowshoe hare doesn't really walk or hop or even leap over the snow. It floats. So does its chief predator, the lynx, which relies on nearly circular, well-furred paws that are huge in proportion the rest of its body. From the perspective of diet, the body of a lynx consists mainly of reconstituted snowshoe hare. As naturalist Ernest Thompson Seton remarked, "The lynx lives on rabbits, follows the rabbits, thinks rabbits, tastes like rabbits, increases with them, and on their failure, dies of starvation in the unrabbited woods."[61] As good a "rabbit" hunter as the lynx is (hares and true rabbits are physically quite distinct), its chances of winter survival are very much at the mercy of snow conditions. Weighing up to ten times more than a hare, a lynx must stalk undetected to within a few bounds of its prey or risk bogging down during the heat of pursuit.

One long-term field study in northern Alberta showed that at the best of times, when the snow was settled and firm, the odds of a lynx chasing down and snatching a snowshoe hare off its feet were one in four. When the snow was particularly soft, with little bearing strength, the lynx's odds of a making a successful kill plummeted to less than one in ten. More precisely, in 91 out of a hundred chases, the hare got away. With a good fluff to the snow, its jumbo feet nearly always prevailed over the lynx's hunger.

Add the fact that its favorite prey turns snow-white every fall, and you can see why, for a lynx, winter is not exactly what you could call a cakewalk. The hare's ability to doff one coat for another with the changing of seasons gives it an alternative name, the varying hare. Its summer coat is buffy brown tipped with black. In response to autumn's dying light, its coat turns snow-white, beginning with the legs, ears, and face, and gradually spreading to the flanks and back. By late November, a hare can crouch invisibly against a snowdrift, betraying its presence only with its dark eyes and black-tipped ears.

Snow is an ally to the hare in other ways, too. Under a heavy snowload, limber branches and shrubs bend over, bringing their tender growing tips within range of the snowshoe hare's well-toothed mouth. When bent, such bushes provide shelter as well as food by forming snow caves, where hares take refuge during periods of severe cold, at, say, −35°C (−31°F) or below. With an arboreal roof of insulating snow overhead—called *qali* by snow ecologists—much of the hare's body heat is conserved rather than radiating into the infinite heat sink of the sky. Consistent snow buildup through the winter favors survival of snowshoe hares by keeping a virtually unlimited supply of food within reach. With each snowfall the hare is lifted higher to fresh twigs and virgin bark. When all the snow melts and the bent-over shrubs and saplings spring back into the air, the tell-tale signs of hare browsing—razor-clean clippings and cuts—often dangle meters above the ground. To the uninitiated eye, it may seem that the forest is inhabited by hare-like creatures about the size of your average bear.

The caribou is another creature of winter, if there ever was one. Its very name speaks of snow. Go back about four hundred years to when the Micmac natives of eastern Canada introduced this animal to the first French explorers. Although something was lost in the translation, they called it *Xalibu,* meaning "pawer" or "shoveler," referring to one of the caribou's most important winter survival skills: digging down through the snow for food with its remarkable feet.

Caribou hooves are actually shaped like shovels, concave and broad, nothing at all like those of any other species of deer. With a few swift strokes of its forefeet, a caribou can break through the crustiest of snow and dig a deep feeding crater to access buried lichens and shrubs. One study from Labrador reports caribou digging craters more than 120 centimeters (4 feet) deep with a "ram hardness" of up to 536 kilograms (1,180 pounds). What this means is that the snow was so hard you could not stick a knife blade into it "without considerable effort."

The caribou's distinctive hoofprint is created by two wide toes shaped like crescent moons. Their edges are sharp and protrude slightly, allowing the caribou to grip ice and hard snow with ease. In winter, the toes are separated by thick tufts of hair that serve as an insulating cushion against the cold. Above the caribou's hoof is a set of metatarsal bones which bend upward with each step, permitting its "dew claws" to share some of the weight and provide extra flotation over the snow.

As distinctive as the caribou's footprint is the profile of its snout. It is thicker and blunter than that of other deer. Its shape is due to extra spaces in the nose which act as heat exchangers, warming inhaled air before it gets a chance to chill the lungs and cooling exhaled air that may rob heat from the body. And inside that extraordinary nose is the anatomical wherewithal to sniff out energy-rich lichens and grasses buried in up to 70 centimeters (28 inches) of snow.

Short ears, small tails, and compact bodies further conserve heat loss by reducing the amount of overall surface area from which precious heat can radiate. Wrapped around this energy-efficient body—even the nose—is a dense overcoat of fur. Over most body parts, it consists of two layers: a thin crinkly inner layer and a thick outer layer of guard hairs filled with bubbles of air. Together these layers surround the animal with millions of pockets of warm air, trapped both within and between the hairs, making the caribou's coat one of the most amazing thermal insulators in nature.

Thus armed to fight the mortal hazards of winter, how could the caribou be anything but invincible when it comes to snow? But things aren't quite that simple. To make the most of their snowy environment (and avoid the worst), caribou must adapt their behavior to the ever-shifting conditions of the snowcover. No one before or since has shown this primordial link more clearly than snow ecologist Bill Pruitt during his now classic caribou study conducted over the winter of 1957-58. Pruitt's avowed mission was to dispel the "virtual ignorance" among mainstream biologists of the ecological significance of prolonged snowcover on caribou. His chosen study area was a generous slice of the Canadian subarctic roughly the size of Nova Scotia or Maine. Here in the northern boreal forest of Saskatchewan and the Northwest Territories, barren-ground caribou spend the winter grazing within a land of spruce-pine woodlands and countless lakes, many with typically enchanting names like Black Lake, Poorfish Lake, and Kasba Lake.

From the outset Pruitt dismissed conventional snow data parceled out by government meteorologists as "woefully inadequate for the purposes of wildlife work . . . The data for biological purposes apparently will have to be collected by the biologists themselves . . . at least until [meteorologists] have been trained in the ecological approach which is, to temperate-zone people, somewhat esoteric."[62] Always one to practice what he preached, Pruitt did just that, personally gathering snow data from more than one hundred sample sites scattered across his immense study area. At each site he measured over a dozen variables, including snow hardness, density, and thickness, using a variety of snow sampling tools, some of which he fashioned himself. He went so far as to engineer a hooflike device to mimic the pressure applied to snow by the footfall of a trotting caribou.

Wintering in the boreal woods of the Northwest Territories, these caribou expertly read the northern snow to find the path of least resistance.
Tessa Macintosh

Now, Pruitt must be as close to a chionophile as any human can get. But during that long subarctic field season, the elements did throw a few curves at him. Many of his carefully exhumed snow samples blew away in the wind before he could weigh them. He later described this as "a constant worry," especially out on blustery lakes and tundra. The delicate work of measuring and dissecting snow required that he take off his gloves, even at "extreme low temperatures." Consequently he froze his hands more than once, especially while handling the industry-standard hardness gauges made of finger-numbing "chunks of brass."

He eventually solved this problem by wrapping the gauges with some good old duct tape and heavy twine. Pruitt's report says nothing about air sickness, eyeball fatigue, inexplicable depression, or other loathsome maladies associated with being stuffed in a small plane while flying back and forth, hour after hour, doing low-level caribou surveys. Any biology types out there know exactly what I'm talking about. You have to think that this too was a challenge for Pruitt considering that, in one winter, he flew almost

15000 kilometers (9,000 miles), the equivalent of traveling more than a third of the way around the planet.

Pruitt's scientific stamina and his passion for snow prevailed, yielding results that made every frost-nipped finger and queasy air mile worth it. He conclusively showed that the timing, direction, and speed of caribou migration were all intimately tied to the hardness, density, and thickness of snow. The highest caribou concentrations were invariably linked to softer, lighter, and thinner snows. Without straying from hard-core snow science, this finding led Pruitt to postulate that varying snow conditions "herd the caribou about" within good feeding and traveling corridors defined by "fences" of harder, deeper snow. He likened these "nivally suitable pathways" for caribou to the aerial flyways that link prime feeding and staging areas for migrating ducks and geese.[63]

Pruitt is a big fan of Alexander Formozov—so big that, he taught himself Russian just so he could help translate Formozov's seminal work on snow ecology, *Snow Cover as an Integral Factor of the Environment and its Importance in the Ecology of Mammals and Birds.* But on one point Pruitt differed with the Russian snow master. Formozov had classified the Old World reindeer (read "caribou"; it's the same species) as a mere chioneuphore— good at tolerating snow, nothing more. After deeply pondering the ways of caribou on their subarctic snow pastures, Pruitt respectfully recommended an upgrade—even a new name.

> Snow cover through which the caribou wade and from beneath which they gather most of their food exerts a profound influence on their behavior, migration and species survival. . . . Since the New World caribou, because of its migrations, is subjected to snow factors for two-thirds of its annual cycle and because it exhibits behavioral and morphological adaptations to snow, we are justified in classifying Barren Ground caribou as a chionophile. . . . Indeed, the name "snow caribou" would be more suitable for this species.[64]

I once suggested to Pruitt that his 1957-58 caribou study must have been something of a marathon. "Scarcely," he said. For this project, he set up *only* 114 snow sampling stations. During his first sabbatical in northern Finland almost twenty years later—this man is dedicated—he established over four times as many stations while chasing reindeer over the snow. In

1983-84 he returned to Finland to dig up and measure snow from no less than 1,363 stations. He determined that in his study area, the snowcover or api as he is fond of calling it, had an average thickness of about a meter. "So," he told me, "I calculated that I dug through about a kilometer and a half of api that winter. I always tell my graduate students they should learn a trade, so that if ever they cannot get a job teaching or researching, they can always earn a living. I have a trade: professional snow-shoveller."[65]

Beware the Disappearing Gift

For snow-adapted creatures like caribou, one of the trickiest times of year— likely *the* trickiest—is not the ice-cold dead of winter. *Au contraire.* Their odds of survival may well peak during prolonged periods of continuous low temperatures and stable snow conditions. It's when winter starts fraying at the seams that things can get dicey, even for the most ardent chionophiles. At no time is snow's role in supporting life more vividly revealed than during its demise. In sympathy with animals that revel in the snow, Pruitt displays a faded bumper sticker on his office door that reads STAMP OUT SUMMER!

As the aerial fisticuffs between winter and spring begin in earnest, the snowcover goes through all kinds of profound transformations, most of which are bad news for animals dwelling both in or above the snow. Settling, crusting, melting, icing, flooding with water, sloughing away—all these processes in turn alter the snow's hardness, density, moisture content, thermal regimes, gas exchange dynamics, and reflectance properties.

Winter often dies in prolonged fits and starts. The thaw begins, then a cold spell sets in, then more thawing, followed by another cold spell. This hedging of seasons brings to mind the image of winter retreating just beyond the hill, still able to suck warmth from the land in a last defiant gesture. Frequent dips above and below freezing play havoc with snow structure, pitting and icing the surface, collapsing basement layers, and saturating all, from top to bottom, with cold dripping water. Voles, shrews, and other subnivean dwellers don't like this. Although these animals can swim, once doused in ice water they are as susceptible to hypothermia as you or I. Wholesale freezing and thawing of soggy snow, if it drags on for many days, can lock small mammals out of their food stores. When temperatures dive yet again, their mortality rises as they become entombed in rotten ice-sheathed snow or succumb to exposure on open snowless ground. This is the time of year I do most of my noninvasive small mammal studies. I often find a lone creature frozen stiff on the forest floor or curled up in a dour patch of rotten snow like a luckless sailor drifting in a lifeboat.

In the so-called dead of winter, the under-snow world is alive with small mammals enjoying warm, pleasantly humid, and remarkably stable conditions. Trouble is, the same warmth that sustains these animals allows bacteria and other microbes in the upper soil to do their business, in a limited way, all winter. Their exhaust fumes, combined with a winter's worth of mousy breaths, result in a gradual buildup of carbon dioxide under the snow. When this gas is trapped by a series of late-winter crusts, it can reach potentially lethal levels. Fortunately, voles and their kin are able to cue in to rising carbon dioxide levels and respond by gnawing ventilator shafts or "snow chimneys" to the surface. Unfortunately, the same shafts that bring fresh life-sustaining air offer an open invitation to deadly intrusions by owls, foxes, and weasels.

When access to shrubs is denied by lowered snow levels, the snowshoe hare can dine on its own droppings, some of which are fortified in proteins and vitamins by intestinal bacteria.
Lorne Schollar

The tail end of winter can also be hard on that living icon of snow, the snowshoe hare. As snow on the ground collapses and the twig-bending loads of *qali* plop to the ground, many hares die. With its prime food now out of reach, the hare resorts to refection or caecotrophy, an unusual dietary adaptation to say the least. In a pinch, so to speak, they munch on their partially digested droppings (the greenish ones apparently), then send them back through their digestive system to squeeze out the last vestiges of nutrition. Now, is that a last resort or what? Apparently not. On their way out, these droppings are enriched in proteins, B-vitamins and other goodies by cellulose munching bacteria that dwell in a special pouch or caecum in the hare's large intestine. Relying on this trick alone, a hare can squeak by for several days. And get this: if you keep a well-fed snowshoe hare in a cage with a wire mesh bottom, it may slowly starve to death because most of those precious droppings fall through before being gobbled up. In the wild, this nifty food recycling system will eventually shut down if nothing new enters the chute from above. At this point, a snowshoe hare's survival depends on either a good dump of snow—to bring food within reach—or a fast melt—to expose previously inaccessible plants. Barring these, a plodding, puddling melt guarantees that its almost magical prowess over the snow will wane, drastically increasing the odds of it ending up between the jaws of its archpredator, the lynx.

Even Pruitt's beloved "snow caribou" can fall on hard times during winter's drawn-out demise. As the sun gains a surer foothold on the sky and the spring snow develops a brittle, ankle-biting sun crust, caribou get decidedly agitated. Their normally self-assured trotting about changes to peculiar bursts of aimless galloping interspersed with spells of idle dawdling. A cold snap at this time of year can cause wholesale freezing of the semi-melted snow. When this happens, the ever-roving caribou comes to a dead stop, beaten by snow too painful to walk through and too hard to dig for food, even with its hoelike hooves. It survives at such times on half-digested cud and an internal shunting mechanism that can slow its basal metabolic rate by as much as 30 percent. Only after the cold snap falters and the snow softens will the roving resume.

Combine spring rains and cold temperatures with unusually deep snow, and you've spelled disaster for caribou. On Nunavut's Bathurst Island, more than 450 of the endangered Peary caribou recently starved to death because they couldn't dig through a rain-soaked layer of deep snow. Several died in their tracks. Weeks later, biologists found some, stone-dead, still standing with their bloodied legs encased in ice-layered snow.

Clearly the ecological pluses and minuses of snow can vary greatly from place to place, day to day, and, of course, species to species. In an article entitled simply "Snow and Living Things," a team of Ontario biologists summed it like this:

A heavy snowload can be a real killer for trees and shrubs. These supple birch near the treeline are doubled over after a long snowy night.
Lorne Schollar

A snow cover may be a boon or a curse for plants and animals.... Snow cover which facilitates travel one day may lethally restrict travel the next, highlighting the fact that snow is a highly dynamic material. . . . Although extreme situations provide clear illustrations of interactions between snow and plants and animals, the most far reaching effects of a snow cover may well have to do with the simple fact that it is there, that it was there last year, and that it will (or should) be there again next year. Plants and animals inhabiting an area *must* interact and respond to it. . . . In those areas where snow cover is a normal part of the environment it is clear that no animal can operate in isolation from something which may, among other things, cover its food, expose or hide it from predators, control the ambient temperature, light regime and gaseous mixture within which it lives. Similar points could be made for plants.[66]

Boon or curse, the upshot of all this is that, from microbes to moose, snow is much more than a deathlike, two-dimensional shroud on the ground. Rather, it is a multistoried mosaic of micro- and macrohabitats which have widely varying suitabilities for both animals and plants. The decoding of these kinds of complex relationships is what snow ecology—or for that matter, *any* ecology—boils down to.

If you've come with me this far, you easily can accept the notion that cold, inert snow offers some attractive real estate for an astounding variety of life forms. Now let me take this thought even further. Snow doesn't simply *allow* for life to carry on between summers, as if winter were some kind of ecological intermission in the grand swing of the seasons. Snow gives life. And more than this, I'm telling you that snow *is* alive—as alive as the oxygen in your blood or the air up your nose. Even without all these creatures milling about under, in, on, or over the snow, this marvellous substance shares in the very life of the planet. Like us, the snowcover's inner anatomy pulsates with invisible energies. Its crystalline skin vents and collects atmospheric gases in a rhythmic sequence of prolonged breaths. Its ever-shifting frame grows and decays over the course of a winter as dictated by the eternal cycling of water. All these vital functions, akin if not synonymous with life itself, stream through the snowcover, even as other life forms take advantage of its life-giving properties.

Make no mistake, however: snow also takes life or at least can make it pretty trying even for the most stalwart chionophile. Adapt or die. This is, after all, the drumbeat to which all life on earth marches. As a dominant thread in the biosphere's tapestry, this life and death struggle with snow is part of nature's epic story. Unlike the remote worlds of physics, chemistry and other orthodox sciences that help decode this story, the world of snow ecology is, for most of us, as handy as a step beyond the back porch in January. Check it out. You never know who you might find hopping around out there on the snow.

Chapter 4
Thinking Snow

Drifting snow,
why do I sing?

—Ojibwa song

Drifts in the Mind

"Winter gives the bone and sinew to Literature," said the late great naturalist John Burroughs, "summer the tissues and blood. The simplicity of winter has a deep moral." Burroughs never spelled out exactly what that moral is, but he more or less told us where to look for it: in the snow. "All life and action upon the snow have an added emphasis and significance," he added, in his typically weighty tones.[67] Burroughs wrote of his ardent love of snow, extolling its soul-nurturing wonders and delights like so many who look at it with a literary eye. Then there is the other camp of writers who portray snow as some life-sucking dark thing that sets all nature a-trembling.

My reckoning of snow literature is that no other element in nature provokes such wildly contrasting emotions. One poet tells you that snow embodies all that is beauty, purity, and order; another declares that it speaks of gloom, wickedness, and chaos. In one book you'll read that snow showers us with priceless treasures, protects all creatures great and small, and rings of eternal benevolence. Pick up another and be told that snow is nothing more than a vexing refuse, a suffocating shroud, a fearful reminder of all that is ephemeral—including our own puny lives.

From my reading of snow poems and stories, all we can agree on here is that snow is riddled with paradoxes and puzzles, or, as snow sage Bernard Mergen proclaims, that snow is "an enciphered message from the cosmos."[68] I find the way in which writers open and read that message to be absolutely fascinating. Some timidly draw the curtains during a snowstorm and hear the mournful clanging of funeral bells. Falling snow irresistibly draws others outdoors for gleeful experiences that blur the boundaries between heaven and earth. Still others sit quietly at home in their igloos, talking to avalanche spirits.

From this multiflavored snow literature I have pulled out for you some appetizing slices. I did this following a carefully planned method that is unique in the history of bibliographic research. No literary scholar before or after me could ever hope to replicate my results. Drawing on the most powerful search engines my whining ten-year-old Macintosh computer could handle, I conducted a rigorous search of the wide world of words about snow. I considered I had made a "hit" only when my two stringent selection criteria were met. First, I had to like the piece, and second, I had to think you might, too.

I considered a literary work up for grabs as long as snow somehow carried its central meaning—be it a major novel or an obscure poem, a narrative of heroic adventure or a wildly inventive yarn, a personal testimony about glimpsing the transcendent or receiving a snowball in the face. Believing that snow offers perennial insights to anyone with a pen in hand, when shopping around I gave nary a thought to era or culture. Enshrined in the words that follow, a contemporary native shaman from the Great Plains rubs elbows with a long-dead Anglo-Saxon minister from Toronto. Stellar personalities like Ralph Waldo Emerson and Robert Frost share these pages with unsung writers discovered by backwoods publishers. Together, they hold up a mirror which shows how we think about snow.

Like snow itself, these words are randomly conceived, yet when clumped together they create distinct patterns on the landscape of our understanding of snow. Drifts and furrows of thought are sculpted by the prevailing gusts and topography of the writers' minds. These shapes are heaped on buried layers of experience and perceptions, ranging from the urban dweller's chronic irritation with snow to the Inuit hunter's existential dependence upon it. I clumped these poems and stories and snippets together to bring out those shapes which, en masse, run the full spectrum of literary sketches inspired by snow—from reverence to revulsion. Between these poles is a whole lot of unsettled emotional ground, unearthed by writers spellbound by snow's many elusive faces.

Stop, drop, and roll in it! A child's hunger for snow stories reflects her untarnished affection for snow.
Northern News Service

Check out your local library terminal or favorite online book browser and type in the word *snow*. Most of your hits will send you to the kids' section. I tried it and scored close to five hundred titles. Fewer than a quarter of them were aimed at adults. There's a reason for this. Unlike many grownups, most children have an untarnished affection for snow. Their hearts lie way over on the reverence end of the snow appreciation spectrum—though of course they'd never call it that. "Awesome" would do just fine for most kids.

So what did I find? Let's start off with *Snowy Day, Snowy Morning* and *The Snowy Night*. Then there's the animal genre, probably the most popular, with *Snow Dog, The Snow Rabbit, The Snow Fox, The Snow Goose* and *The Penguin in the Snow*. Care for the supernatural? Reach for *The Snow Angel, The Snow Giant, Caverns of the Snow Witch* and, of course, *Frosty the Snowman*. Or how about a mixed bag of snow stories, poems, and jokes? Try *Snow Tales, Snow Stories* and *Snow Stumpers*. There is also a growing list of titles aimed at getting young people out in the snow to study its many moods and properties: *Snow Watch, The Secret Language of Snow* and simply *Snow*—I found several children's books with this irresistible one-word title.

Snow angels, snow giants, snow princes—children's literature is heavily peopled with such radiant beings. Yellowknife's self-proclaimed "Snow King" annually rides this wave of popular fantasy with a children's theater in his snow palace.
Fran Hurcomb

I expect that young snow lovers are too busy reading these kinds of things to pick up my book. So my focus here is on adult material—though I assure you this boundary is not sharp. But allow me a quick diversion into juvenile snow literature with one poem and one story—two of my favorites.

"First Snow" by Marie Louise Allen basically captures all that need be said about snow: it falls on all without prejudice, it conjures beauty wherever it lies, and it transports us to mysterious playful worlds.

> *Snow makes whiteness where it falls.*
> *The bushes look like popcorn-balls.*
> *And places where I always play,*
> *Look like somewhere else today.*[69]

My favorite snow story off the kids' shelf is *The Friendly Snowflake* by Scott Peck, of *The Road Less Traveled* fame. This wonderfully wrought fable begins with a lone snow crystal landing on heroine Jenny's nose, sparking her voyage into some pretty heavy but delightfully explored spiritual questions. What is life, love, and faith?—that sort of thing. These questions come into

"The bushes look like popcorn-balls. . . ." —Marie Louise Allen
Lorne Schollar

play as Jenny and her agnostic brother Dennis explore the life history of a snow crystal. Jenny gets the metaphysical snowball rolling by quizzing her brother—a modern-day Job—on how many dump trucks it would take to blanket their modest backyard with 2 feet of snow. Dennis jumps right into her trap. After some head and paper scratching, he looks up with a big grin and announces that it would take 87.

> "Then how many dump trucks would it take to blanket all of
> New England with two feet of snow?" Jenny asked.[70]

After diving for his atlas and another round of scribbling and scratching, Dennis comes up with a nice round figure of 1.75 billion. And here's where Jenny cajoles Dennis to hand over the credit for a task of such cosmic proportions to the proper authority.

> "Are there that many trucks in the world?" Jenny wondered.
> Dennis thought for a moment. "Nope. Not nearly."
> > "So it would take more trucks than there are in the whole
> > world just to bring this one storm to our little part of the coun-
> > try," Jenny mused. "God sure must be pretty powerful."

Put yourself in the smallest kid's shoes: you're six years old, you just immigrated to northern Canada from some tropical place, and on your first day there it starts snowing. To snow's natural appeal for kids, add the fact that you've never even seen it before, and you're in for one exciting morning—as it was for this newly arrived family from the Philippines.

Northern News Service

This does not sit well with Dennis, who tries to douse her explanation with classroom meteorology. That night Jenny thanks God for bringing such a big storm, adding that it must have been "a lot of work." Though this may be a little too cute for some tastes, what charms me about this story is that Peck takes us beyond the well-trod wonders of snow to the wondermaker him-/her-/it-self (take your pick), bolstering the bald-faced fact that snow really is quite miraculous.

Plunging into Winter

You've probably heard the one about how you can tell a lot about a person by the way he or she deals with flat tires, crying babies, and snarled Christmas lights. Speaking from experience, I would hasten to add tangled coat hangers. But even more telling, to my mind, is the way people come to terms with winter's first good dump of snow. Remember how, as kids, we greeted those first flakes: with open mouths and stuck out tongues. When was the last time you did *that?* I suspect that if Minnesota nature writer and snow inspector Sigurd Olson was alive, he'd still be at it. In *The Singing Wilderness,* he cheerfully welcomed "those first drifting flakes" as a call to the simpler, more laid-back pace of winter for "both man and beast."

> To me, that is the real meaning of the first snowfall—not the ces-
> sation of effort, but a drawing of the curtain on so many of the
> warm-weather activities that consume so much time. The snow
> means a return to a world of order, peace, and simplicity. Those
> first drifting flakes are a benediction and the day on which they
> come is different from any other in the year.[71]

However you welcome those first flakes, with protruding tongues or slammed doors, the coming of snow signals a stark irrevocable transition, giving us pause to take stock and prepare for darker, colder times ahead. If you live in the Yukon, as did freelance journalist John Dunn when he wrote "The Countdown from Light to Dark," this plunge into winter is especially steep.

> While the true outside is realized in winter, so is the real inside.
> There is anonymity in darkness, a feeling of isolation. The snow
> muffles the sounds of traffic and the world is seen through the
> narrow tunnel of a parka hood. The onset of winter in the Yukon
> is like a suspended time of reckoning; long after the leaves have

left the trees, the snow works its way down the mountains, the new season's calling card. The warning gives us time to pause.[72]

The onset of winter darkness: "a suspended time of reckoning." –John Dunn
John Poirier

The world of New England's poet laureate Robert Frost revolved around birches, crows, and snows. His poems of winter's whims and handiwork are so vivid that you'd swear he snatched them on the fly, his long artist's fingers scribbling in the wind. In an early winter poem, "Wind and Window Flower," Frost draws a delicate boundary between the inside and outside of winter by weaving an unlikely love story between an icy wind and a frail flower framed by a "warm stove-window." The winter wind "sighed upon the sill / He gave the sash a shake. . . ." But fickle flower gave him the cold shoulder. Winter was already too far gone for the likes of her to venture out.

> *He was a winter wind,*
> *Concerned with ice and snow,*
> *Dead weeds and unmated birds,*
> *And little of love could know.*
> *. . . But the flower leaned aside*
> *And thought of naught to say,*
> *And morning found the breeze*
> *A hundred miles away.*[73]

As spring is borne on the wings of birds and summer on rays of the sun, winter, more than any season, is borne on the wind. Its central place in written works featuring snow comes as no surprise. While riding through the feral forests and mountain valleys of the Yukon, C. J. Pettigrew has watched closely the lashings of winter winds on her horses. There's nothing romantic about her portrayal of winter's rougher edges. What I read here is prudent backwoods acceptance about the way things just plain are. Between the lines she's saying, if you want to go riding into a Yukon snowstorm, make sure you take your toughest horse—and an extra bag of feed. That wind-driven snow can suck the life out of it.

> There is no magic in horses in the winter. The long clean limbs that delighted us in summer now grow shaggy and blunt. Blurry equine profiles hang low; their lips dangle over frozen clumps of milkweed. They are motionless except to shift their weight from one arthritic knee to another. The old dead hairs of their tails mingle with the dead grasses.
>
> When the wind lashes down from the high country, salted with crystals of ice, the young horses snort and roll their eyes. The old horses stretch out their necks to the bony hand of winter.
>
> It takes 4.8 calories of body heat to melt a mouthful of snow for water, and it takes a gallon of water to digest a bellyful of fodder. Every degree of cold exacts more water, body heat, more food. But the cruellest exactor is the wind: only a doomed horse will leave the shelter of bare trees to graze with the wolfish wind.[74]

Shroud of Darkness and Death

I figure they need more snow over in Britain so they can get used to it, maybe even befriend it. Death and snow seem to be forever linked in the minds of their wordsmiths. If there was an award for painting the bleakest word pictures of snow, the British would steal it every time. Take the final lines of James Joyce's *Dubliners:* "His soul swooned slowly as he heard the snow falling faintly through the universe and faintly falling, like the descent of their last end, upon all the living and the dead."[75] Hardly cheery stuff. Or how about Henry Wadsworth Longfellow? His poems speak of snow as a "countenance [of] confession," a tangible byproduct of despairing and grief-stricken skies. In "Afternoon In February" he pulls all stops, painting snow as the perfect backdrop for all things dead and dying.

An abandoned tent frame stands up to ominous winter skies and a wind "colder than death." What snowy poems might the likes of Longfellow have written in such a place?
Lorne Schollar

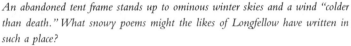

> *The day is ending*
> *The night is descending;*
> *The marsh is frozen,*
> *The river dead.*
> *Through clouds like ashes*
> *The red sun flashes*
> *On village windows*
> *That glimmer red.*

Okay, if you thought things were grim with ashen clouds and blood–red sunsets, Henry now conjures up a ponderous snowfall that obliterates all that is familiar and paves the way for a funeral parade.

> *The snow recommences;*
> *The buried fences*
> *Mark no longer*

The road o'er the plain;
While through the meadows,
Like fearful shadows,
Slowly passes
A funeral train.
The bell is pealing,
And every feeling
Within me responds
To the dismal knell;
Shadows are trailing,
My heart is bewailing
And tolling within
Like a funeral bell.

The Dead Poets Society of Britain has no monopoly on spreading the idea that snow means death. Commenting on the many faces of snow in American art, 1920s journalist Benjamin Flower described "the great Mother [as] mantled in her shroud of snow . . . sombre and silent in the recuperative sleep that so resembles death."[76] In "December Storm," Canadian poet Ashton Green portrays snow as "deceitful" and winter's icy touch as "colder than death." But he moves beyond Longfellow's despairing imagery, advising us to open our eyes to winter's hard knocks, however bracing, and trust in their lessons.

Deceitful snow, made of the gray sea,
Half water and half ice—the wind
Screams we must have it, and a rage
Of indecision blinds the world,
Colder than death; but it will end.
We will look up and see again;
Having been whipped across the eyes
Like penitents, to make us see.[77]

Robert Service craftily exploits the image of snow as a backdrop of doom in many of his poems, some ribald and wild, others deadly serious. In "Sunshine," the poet's beloved wife lies stone-cold dead in a remote cabin "on the wind-flailed Arctic waste." Enveloped by a forbidding snowscape with "silence, darkness, death around," he awaits the miracle of her return— or begs a homeward journey to her through his own death.

"Flat as a drum-head stretch the haggard snows . . . on the wind-flailed Arctic waste." –Robert Service.

Lorne Schollar

> *Flat as a drum-head stretch the haggard snows;*
> *The mighty skies are palisades of light;*
> *The stars are blurred; the silence grows and grows;*
> *Vaster and vaster vaults the icy night.*
> *Here in my sleeping bag I cower and pray:*
> *"Silence and night have pity! Stoop and slay."*
> *The cabin must be cold, and so I try*
> *To bear the frost, the frost that fights decay,*
> *The frost that keeps her beautiful alway.*
>
> . . . .
>
> *She lies within an icy vault;*
> *It glitters like a cave of salt.*
> *All marble-pure and angel-sweet*
> *With candles at her head and feet.*[78]

In *Beautiful Snow*, James Watson explores the dark world of snow by holding up its purity and radiant beauty as a mirror for a street-dweller's inner ugliness and pain. Watson takes us on a precipitous fall into darkness—I never saw it coming the first time through. In the end, his woeful character is pushed to the brink of suicide by a town "gone mad in its joy at the snow's coming down."

> O the snow, the beautiful snow,
> Filling the sky and the earth below.
> Over the house-tops, over the street,
> Over the heads of the people you meet,
> Dancing,
> Flirting,
> Skimming along,
> Beautiful snow, it can do nothing wrong . . .
> It lights up the face and it sparkles the eye;
> And even the dogs, with a bark and a bound,
> Snap at the crystals that eddy around.
> The town is alive, and its heart in a glow
> To welcome the coming of beautiful snow.
> How the wild crowd goes swaying along,
> Hailing each other with humor and song!
> How the gay sledges like meteors flash by,
> Bright for a moment, then lost to the eye!
> Ringing,
> Swinging,
> Dashing they go,
> Over the crest of the beautiful snow.
> Snow so pure when it falls from the sky,
> To be trampled in mud by the crowd rushing by:
> To be trampled and tracked by the thousands of feet
> Till it blends with the horrible filth in the street.[79]

Did you catch that schizophrenic flip-flop? Between breaths, Watson hurls us from one end of the snow spectrum to the other, from purity to purulence. With that two-faced mirror thrust before him, the street-dweller is about to fall headlong down a very slippery slope—with still a long way to rock-bottom.

> Once I was pure as the snow—but I fell,
> Fell, like the snow-flakes, from heaven—to hell;
> Fell, to be tramped as the filth of the street,

Fell, to be scoffed, to be spit on, and beat,
Pleading,
Cursing,
Dreading to die,
Selling my soul to whoever would buy;
Dealing in shame for a morsel of bread.
Hating the living and fearing the dead.
Merciful God! Have I fallen so low?
And yet I was once like the beautiful snow!
And once I was fair as the beautiful snow,
With an eye like its crystals, a heart like its glow;
Once I was loved for my innocent grace—
Flattered and sought for the charm of my face.
Father,
Mother,
Sisters, all,
God, and myself, I have lost by my fall . . .
How strange it would be when the night comes again,
If the snow and the ice struck my desperate brain,
Fainting,
Freezing,
Dying alone,
Too wicked for prayer, too weak for a moan
To be heard in the crash of the crazy town,
Gone mad in its joy at the snow's coming down;
To lie and to die in my terrible woe,
With a bed and a shroud of the beautiful snow.

Ouch. Nowhere else have I found such a vivid use of snow as a metaphor for human misery. In my hip-deep wadings through wintry literature, I also discovered that snow is often used as a stage for portraying the misery of animals. In Morley Callaghan's *They Shall Inherit the Earth,* for instance, I learned that "there is nothing quite so dead" as a deer "stuck stiff in the snow." Set against Canada's sprawling north woods, his novel leaves you guessing as to who is more villainous—the brutish hunter or the deep virgin snow.

They crossed the deer trails and were out on the hard crust, and when they had gone a little way they saw that the mound was really the carcass of a deer. There were many other carcasses, and blood marks, with the snow churned up, and the carcasses stuck there in the hard crust; the sharp hoofs of the deer, piercing through the crust, had impaled them in the snow. A deer stuck stiff in the snow is very dead; there is nothing quite so dead. The carcasses were slashed at the throat, or slashed at the nose, and the flesh of the tenderloin had been torn from every one of the carcasses, just the nice, juicy tenderloin torn away, and the rest of the carcass left there to bleed and freeze in the snow.[80]

Blood on the snow. Can you think of a more shocking image in nature? I get the primal jitters when, sketched across the snow, I see the crimson remains of a hare ripped apart by a great horned owl, or a ring of red around the hoof mark of a moose that had struggled through a sharp spring crust.

As a soothing balm to the menacing face of snow, allow me to introduce you to American botanist William Gibson, whose writings illustrate the flip-side of the "suffocating shroud" view of snow. He looks at it as a benevolent blanket that smoothes nature's rough edges and levels society's warts. In an 1885 issue of *Harper's* magazine, Gibson took this sentiment to rhetorical extremes with "A Winter Walk."

[H]ow this clear, purged atmosphere sharpens the sight and opens the horizon, as the merciful mantle of snow smoothes away all former invidious distinctions, and confounds our arbitrary judgements! In these white fields you shall not know poverty from affluence, worldly distinction from obscure humility. The princely park and the plebian potato patch are one; their artificial barrier is blotted out in this universal baptism of beneficent whiteness.[81]

Braving the Storm

So far, a menagerie of writers and poets has transported you well into winter, and you've got both poles of the snow appreciation spectrum in full view. Now here come the snowstorms, which elicit a wide array of emotions, most of them having everything to do with high-spirited conflict.

A writing mentor once told me that there are three kinds of conflicts around which the best poems and stories revolve: Man Against Man, Man Against Nature, and Man Against Himself. (Sorry, ladies, but that's the way I learned it.) If this is true—and I believe it is—a well-written tale about a rip-snorting snowball fight in a grand mal snowstorm would probably be a shoe-in for the Pulitzer Prize. There's nothing like a good snowstorm to bring out winter's claws, reminding us of the thin veil that shields us from nature's life-snuffing power. In *Storm Fear,* even a snow lover like Robert Frost finds that veil wearing thin when barking winds and pelting snow pound at his farm-house door.

> *When the wind works against us in the dark,*
> *And pelts with snow*
> *The lower chamber window on the east,*
> *And whispers with a sort of stifled bark,*
> *The beast,*
> *"Come out! Come out!"*—
> *It costs no inward struggle not to go,*
> *Ah, no!*
> *I count our strength,*
> *Two and a child,*
> *Those of us not asleep subdued to mark*
> *How the cold creeps as the fire dies at length*
> *How drifts are piled,*
> *Dooryard and road ungraded,*
> *Till even the comforting barn grows far away,*
> *And my heart owns a doubt*
> *Whether 'tis in us to arise with day*
> *And save ourselves unaided.*[82]

At first glance you might think Frederick Philip Grove's *Over Prairie Trails*[83] was a bit of a yawner. Here's a folksy 1922 book of nature essays about Manitoba's changing seasons written by a teacher on his way to and from his schoolhouse. But crack it open, turn to the essay called "Snow," and buckle up for one of the most nail-biting, heart-flipping, jaw-dropping snow stories you'll ever read. The narrative opens with a "blinding northern gale," which throws up "mountains" of snow between the author's schoolhouse and his wife and home over 50 kilometers (30 miles) away. The "unfailing temerity" with which he confronts the storm and its awesome handiwork is a gripping study of man against himself as much as against the elements. In a horse-drawn sleigh which "altogether behaved like a boat tossed on a stormy sea," he crosses what must be the biggest snowdrifts ever immortalized in ink. Besides a few moments of bafflement when, for instance, he faces what seemed to be "all the snow of the universe," Grove remains cool as a cucumber. Keep in mind that this is not fiction we're dealing with here. The guy actually went out in that storm-whipped snow.

> The blizzard started on Wednesday morning. It was that rather common, truly western combination of a heavy snowstorm with a blinding northern gale—such as piles the snow in hills and mountains and makes walking next to impossible . . . When I stepped outside again, the wind seemed bent on shaking the strongest faith.

It turned out to be a three-day blow, one of those prairie "orgies in which Titan Wind indulges every so often." Right up front, Grove casually remarks that he has been lost in a blizzard more than once. I get the strong impression that he quite enjoyed the experience; at the beginning of his story, he's telling us it's high time to get lost in another one. He expresses unconcealed disdain for "the inhabitant of the middle latitudes" who knows nothing of such storms and views snow as "something that suggests the sleep of Nature rather than its battles." On the eve of his proposed journey home, his friends ply him with tea, cakes, and polite pleas to desist.

"Do not try to dissuade me," he says resolutely. "I am sorry to say but it is useless. I am bound to go." He sets out the next day before dawn, primed to do battle with the snow. "As long as there was the least chance that horse-power and human will-power combined might pull me through at all," he writes, "I was determined to make or anyway to try it."

> It was still dark, of course, when I left the house on Saturday morning to be on my way. Also, it was bitterly cold, but there was very little wind. In crossing the bridge which was swept nearly clean of snow I noticed a small, but somehow ominous-looking drift at the southern end. It had such a disturbed, lashed-up appearance. The snow was still loose, yet packed just hard enough to have a certain degree of toughness. You could no longer swing

your foot through it: had you run into it at any great speed, you would have fallen; but as yet it was not hard enough to carry you. I knew that kind of drift; it is treacherous. On a later drive one just like it, only built on a vastly larger scale, was to lead to the first of a series of little accidents which finally shattered my nerve. That was the only time that my temerity failed me.

With uncanny acuity, this man sees more details in plain old snow than most of us would discover in a painting by Hieronymus Bosch. My conclusion is that Grove must have been an Inuk in a former life. He uses his discriminating eye for the nuances of snow to tease out its many moods: "inhospitable, merciless, and cruelly playful." After rendering snow's features so graphically that you'd swear, as you read, your own cheeks are stinging, he moves from quality to quantity, describing volumes of snow capable of swallowing whole landscapes.

"Altogether there was an impression of barren, wild, bitter-cold windiness about the aspect that did not fail to awe my mind: it looked inhospitable, merciless, and cruelly playful . . ." —Frederick Philip Grove.
A late winter storm takes flight over bald tundra near Iqaluit, Nunavut Territory.
Tessa Macintosh

We turned at a walk, and the chasm of the bush road opened up. Instantly I pulled the horses in. What I saw baffled me for a moment so completely that I just sat there and gasped. There was no road. The trees to both sides were not so overly high, but the snow had piled in level with their tops; the drift looked like a gigantic barricade . . . Here I felt for a moment as if all the snow of the universe had piled in.

Don't you just love that? "All the snow in the universe." Who but a snow inspector like Frederick Philip Grove could get away with a hyperbole like that? And here comes the dénouement: horsepower and willpower unite in victory as Grove skippers his sleigh to heights surpassed only by Santa Claus.

The snow proved harder than I had anticipated—which bespoke the fury of the blow that had piled it. It did not carry the horses, but neither—once we had reached a height of five or six feet—did they sink beyond their bellies and out of sight. I had no eye for anything except them. . . . They went in bounds, working beautifully together. Rhythmically they reared, and rhythmically they plunged. I had dropped back to the seat, holding them with a firm hand, feet braced against the dashboard; and whenever they got ready to rear, I called to them in a low and quiet voice, "Peter—Dan—now!" And their muscles played with the effort of desperation. It probably did not take more than five minutes, maybe considerably less, before we had reached the top, but to me it seemed like hours of nearly fruitless endeavour. I did not realize at first that we were high. I shall never forget the weird kind of astonishment when the fact came home to me that what snapped and crackled in the snow under the horses' hoofs, were the tops of trees.

Grove makes it home all right, announcing breezily to his wife, "I have seen sights today that I did not expect to see before my dying day." For him, the snowstorm set the stage for a first-rate man against nature story in which an iron will and a solid grounding in the language of snow sees the hero through to a benevolent truce with nature. For Loren Eiseley, in "The Angry Winter," a snowstorm set in a cemetery creates a darker theater in which death and life swirl in precarious balance. Having told the epic story of how young *Homo sapiens* managed to walk all the wiser out of the last ice age,

Eiseley confronts his own mortality reflected in the eyes of a hungry cowering jackrabbit—a fellow wraith, "fading in the storm."

> Many years ago, as a solitary youth much given to wandering, I
> set forth on a sullen November day for a long walk that would
> end among the fallen stones of a forgotten pioneer cemetery on
> the High Plains. The weather was threatening, and only an
> unusual restlessness drove me into the endeavour. Snow was on
> the ground and deepening by the hour. There was a rising wind
> of blizzard proportions sweeping across the land.[84]

Like Grove, Eiseley is the kind of fellow magnetically drawn outside
by a good snowstorm. He never says exactly why he set his sights on an
old broken-down cemetery, but it makes for a good story—this one also
true. Of course he reaches the cemetery just before nightfall. As snow fills
the air, he beholds a mess of cracked and toppled tombstones. Sneaking
up on him is the feeling that he is "the last living man . . . freezing among
the dead."

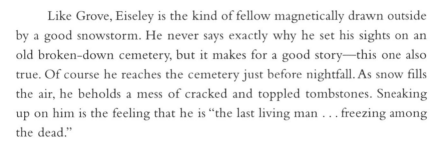

> I leaned across a post and wiped the snow from my eyes. . . . It
> was then I saw him—the only other living thing in that bleak
> countryside. . . . He was nothing more than a western jack rab-
> bit, and his ribs were gaunt with hunger beneath his skin. Only
> the storm contained us equally. That shrinking, long-eared ani-
> mal, cowering beside a slab in an abandoned graveyard, helpless-
> ly expected the flash of momentary death, but it did not run.
> And I, with the rifle so frequently carried in that day and time,
> I also stood while the storm—a real blizzard now—raged over
> and between us, but I did not fire.

What unites hunter and prey is not the latent power in Eiseley's trigger
finger but the raging snowstorm that could bury them both without a trace.
But, like our hairy relatives of yore, Eiseley walks out of the storm unscathed
and a little wiser. We never learn what happens to his bunny friend.

> Step by step I drew back among the dead and their fallen stones.
> Somewhere, if I could follow the fence lines, there would be a
> fire for me. For a moment I could see his ears nervously record-
> ing my movements, but I was a wraith now, fading in the storm.

"There are so few tracks in all this snow," someone had once protested. It was true. I stood in the falling flakes and pondered it. Even my own tracks were filling. But out of such desolation had arisen man, the desolate. In essence, he is a belated phantom of the angry winter. He carried, and perhaps will always carry, its cruelty and its springtime in his heart.

Some storms, however charming, simply won't allow us to venture outside, as Monte Hummel will tell you. As president of World Wildlife Fund Canada, Hummel is, to say the least, a busy man. His job of convincing *Homo industrialis* to curb his juggernaut and stop trouncing on wild nature leaves little time for smelling the very flowers he's trying to protect. But what good are such captains of conservation to their cause if they don't escape to the bush now and then to confer with their primary client? Not much, according to Hummel, who regularly recharges at his remote backwoods cabin nestled on the shore of a small shield lake in northeastern Ontario. He even managed to write a book out there, some of it by candlelight. In *Wintergreen: Reflections from Loon Lake,* Hummel describes "a howling winter whiteout" that engulfed his cabin for two days, forcing him to climb out a window after the storm finally blew itself out. What's especially poignant here is the arresting transition from sinister storm to a benevolent "gloaming," both products of "the same winter's force."

> Without fail, a freshening breeze with wet gray-blue clouds on the eastern horizon at Loon Lake means "Get ready for something interesting." . . . By late morning, I returned from my walk through heavy snowflakes that were gently descending vertically. By noon they were slanting westward. And by midafternoon, my cabin was engulfed in a howling winter whiteout.
>
> I've experienced storms in all seasons—out on lakes and rivers, deep in the bush, inside a tent, even under the shelter of an overturned canoe tipped up against the wind. But there is something particularly satisfying about gazing out on winter's rage from the warmth and safety of a secure, warm cabin. The stove may back-puff in the gusts, the candle flame may flutter at night, but I know I'm basically well provided for and left with only a few simple things to do—sit tight, keep the fire burning, and wait the storm out.[85]

From what I've read—and experienced at my own cabin—I can confidently tell you that there is an inverse relationship between the sharpness of a snowstorm's teeth and the coziness of your favorite comfy chair. At this point in the story, Hummel pulls his chair closer to the woodstove and imagines sparks "being ripped horizontally downwind into the dark, along with the odd shingle." When he falls asleep by the fire, the wind blows on through his dreams. Meanwhile the storm spins an impenetrable cocoon of snow around his cabin.

> By noon on the second day, I wonder whether this storm is even capable of fatigue. It roars its answer defiantly through the screened porch, driving more and more snow over the firewood I have stacked there. Just when I think I sense some weakening, a steady deep-throated power rolls across the land and brings the full force of Canadian winter to bear on my tiny shelter.
>
> The wind totally obliterates the horizon, unifying land and sky. Now there is no telling whether the blended mass I see out my window is fallen snow lifted up into the air, or new snow driven down from above. For a moment, I feel genuinely threatened by the deafening wild violence all around me, and I wonder whether my cabin will be able to withstand its force.

Then, finally, it comes: "a hard-won silence"—the kind of silence that makes prairie dwellers bolt upright in their beds, wide awake, when the wind abruptly dies in the middle of the night. Hummel breaks out of his cocoon to admire a land recast and cleansed by the storm.

> My cabin door is blocked by snow on the outside, so I clamber out a window, step into a thigh-deep snowdrift, liberate a shovel by hand-digging under the floorboards, then dig out my entrance. I also clear a place on the front stoop, just to sit and take in the winter storm's aftermath, a cup of fresh steaming coffee in hand . . .
>
> Before me is a world transformed, so beautifully clean and sculpted that I seriously weigh whether it is right to mar its surface with my tracks. In the end, I conclude that if the grouse and otter have ventured out from their shelters, as evidenced by their footprints, then so may I. The same winter's force that held me

cabin-bound for two days has now given way to a soft gloaming, and reshaped a land waiting to be reanimated by all of us who live here.

For a moment, Hummel is paralyzed by the stunning art gallery thrown open by the storm. His wilderness domain is adorned with lithe and lovely snow sculptures wrought by the wind. Ralph Waldo Emerson explores this playful, shape-shifting power of snow after a wild New England tempest. "The Snow-Storm" abruptly takes off with an aerial trumpet blast and wind-driven snow. For a few brief lines we are, like Hummel in his cabin, snugly sealed off in "a tumultuous privacy of storm." Then Emerson boldly invites us outside to admire "the mad wind's nightwork," which makes a merry mockery of anything we humans could construct.

> Announced by all the trumpets of the sky,
> Arrives the snow, and, driving o'er the fields,
> Seems nowhere to alight: the whited air
> Hides hills and woods, the river, and the heaven,
> And veils the farm-house at the garden's end.
> The sled and traveler stopped, the courier's feet
> Delayed, all friends shut out, the housemates sit
> Around the radiant fireplace, enclosed
> In a tumultuous privacy of storm.[86]

Here Emerson takes us down a familiar road: the raucous snowstorm, its sudden demise, the busting outside to revel in its handiwork. But, deferring to "the unseen," he refuses to pigeonhole this storm as either good or evil. Instead he paints its face with many colors—fierce, savage, and mad; fanciful, frolicsome, and free—so that in the end, it transcends any labels we may try to stick on it. What else would you expect from one of America's greatest transcendentalists?

> Come see the north wind's masonry.
> Out of an unseen quarry evermore
> Furnished with tile, the fierce artificer
> Curves his white bastions with projected roof
> Round every windward stake, or tree, or door.
> Speeding, the myriad-handed, his wild work
> So fanciful, so savage, naught cares he

For number or proportion. Mockingly,
On coop or kennel he hangs Parian wreaths;
A swan-like form invests the hidden thorn;
Fills up the farmer's lane from wall to wall,
Maugre the farmer's sighs; and, at the gate,
A tapering turret overtops the work.
And when his hours are numbered, and the world
Is all his own, retiring, as he were not,
Leaves, when the sun appears, astonished Art
To mimic in slow structures, stone by stone,
Built in an age, the mad wind's nightwork,
The frolic architecture of the snow.

"The mad wind's nightwork, the frolic architecture of the snow . . . " —Ralph Waldo Emerson.

Swatches of windswept snow accent ice blocks trapped on a freshly frozen lake.
John Poirier

Many-Splendored Snow

Like searing deserts, towering mountains, pounding surf, and other evocative landscapes, a fresh dump of snow, wherever it lies, seems to leave a common signature on the minds of many writers. In *The Perfection of the Morning*, Sharon Butala reflects on snow's imprint on the ranchland of southwestern Saskatchewan and on her heart. She hits buttons well known to us all when she writes of snow's call to innocent pleasures, of its universal power to transform the familiar into the foreign and to spark nostalgic memories of our youth. Describing her early days in ranch country, she remembers her first walks each morning along the snow-filled riverbed "as among the best of my life."

> Looking up from the snowy riverbed, I saw white walls of snow and then the snowy billows and beyond them the brilliant sky. I saw the places where partridges snuggled up for the night to keep warm and followed the tracks of coyotes and foxes and animals whose tracks I didn't recognize. I was picking up knowledge, hardly even noticing that was what I was doing. Running to cut off a cow, I fell headlong in the snow and, with no one watching me, lay there laughing, blinking up at the sky, losing myself in its blue depths.[87]

As a neophyte wife and ranchwoman, one of the first Christmas gifts Butala received from her husband was a pair of cross-country skis. As long as there was enough snow, which is hardly the norm in that parched corner of the prairies, she would be out there on her skis, trucking off into the hills. Whenever possible, she made a habit of skiing into snowstorms, having discovered that she liked storms "for the way they changed the appearance of familiar places and for the sense of mystery they brought to them." These kinds of experiences cast her thoughts back to earlier days filled with snow.

> Memories of my childhood came back to me: playing in the bush with my friends, with my sisters and cousins in our grandmother's garden, skating on frozen sloughs in winter till the pain from the cold became so bad even we kids couldn't stand it anymore and went home, the winter we had built a snow fort that lasted for months as we added on and made it more and more substantial so that it stood well into spring. I felt like a child again, had fleeting moments when I remembered how wonderful the world itself had once seemed, and how it was to be cared for, worry-free, and living in the body again and not just the mind.

What is this magic in snow that can transport us back to "worry-free" snatches of our childhood? In "Snow by Morning," May Swenson starts her poem with the promise of snow "for everyone," which in the end brings the gift of regained youth.

> *By morning we'll be children*
> *feeding on manna,*
> *a new loaf on every doorsill.*[88]

So powerful is snow's magic that even the mere memory of playing in
it can put us in a harmonious mood. In *No Ordinary Time,* Doris Kearns
describes how snow once held a pivotal role in maintaining order, even on
the stage of presidential politics.

> On nights filled with tension and concern, Franklin Roosevelt
> performed a ritual that helped him to fall asleep. He would close
> his eyes and imagine himself at Hyde Park as a boy, standing with
> his sled in the snow atop the steep hill that stretched from the
> south porch of his home to the wooded bluffs of the Hudson
> River far below. As he accelerated down the hill, he maneuvered
> each familiar curve with perfect skill until he reached the bot-
> tom, whereupon, pulling his sled behind him, he started slowly
> back up until he reached the top, where he would once more
> begin his descent. Again and again he replayed this remembered
> scene in his mind.... Thus ... the president of the United States
> would fall asleep.[89]

Christmas and snow—now, *there* is a writer's recipe for pure aesthetic
delight. Add to this mix a cozy cabin in a wilderness setting, and you've got
me leaking good cheer from all pores. These enticing streams all come
together in Theodora Stanwell-Fletcher's *Driftwood Valley,* where she tells the
story of how she and her husband lived in "an unexplored wilderness" in
north-central British Columbia in the late 1930s, collecting flora and fauna
for the provincial museum in Victoria. Holding degrees in both literature
and animal ecology, she put her dual training to particularly good use in this
colorful chronicle of Christmas Day 1938.

> By the time dawn was coming we had scraped two peepholes in
> the frost on the panes; and we stood quiet to watch the winter
> sunrise. The radiant peaks of the Driftwoods, cut like white icing
> into pinnacles and rims against the apple-green sky, were
> brushed with pink, that, even as we watched, spread down and
> down and turned to gold. Rays of the rising sun, coming
> between the pointed firs of the east shore, stretched straight

across the white lake, and as they touched it huge crystals, formed by the intense cold, burst into sparkling, scintillating light. The snow-bowed trees of the south and west shores were hung with diamonds; and finally the willows, around our cabin, were decked with jewels as large as robins' eggs that flashed red and green and blue. No Christmas trees decorated by human hands were ever so exquisite as the frosted trees of this northern forest. The sky turned to deep, deep blue, and the white world burst into dazzling, dancing colors as the sun topped the forest. The dippers, undismayed by a cold that froze dumb all other living things, broke into their joyous tinkling melody by the open water patch below the bank. And our first Christmas Day in the wilderness was upon us.[90]

Stanwell-Fletcher took quite a liking to those dauntless dippers, drab gray birds with a habit of walking underwater on the bottom of rushing mountain streams. In spite of her credentials, she never could figure out how they went about their bird business so "completely unaffected" by arctic-like temperatures. Twice during December alone, hand in hand with her husband, I imagine, she watched a pair of dippers merrily mating on a snowbank.

By noon things had moderated enough for them to go out for a rousing "tramp" on their snowshoes. By "moderated," Stanwell-Fletcher means that instead of −45°C (−50°F), the thermometer had leaped up to a blazing −38°C (−36°F).

Our snowshoes tossed up clouds of crystals. Young trees which, in autumn, had reached above our heads had been completely covered with fresh snow, so that they were transformed into great mounds and small hills. Wherever we looked our eyes were dimmed by the twinkling brilliants scattered before us. The azure of the sky above, the unsullied whiteness below, the mountains and the woods, the intense pureness of the air, were exhilarating beyond imagining. And there was not a sound or a motion, anywhere, to distract our senses of sight and feeling.

As daylight faded, the happy couple was treated to red and lavender snows and purple-topped mountains. Then something marvelous happened, what Stanwell-Fletcher calls "the most wonderful" of all the day's spectacles.

It is that moment of white twilight which comes on a particularly clear afternoon, after the last colors of sunset fade and just before the first stars shine out. I don't suppose its like can be seen anywhere except in the snowbound, ice-cold arctic places. Everything in the universe becomes a luminous white. Even the dark trees of the forest, and the sky overhead, are completely colorless. It is the ultimate perfection of purity and peace.

From the interior mountains of northern British Columbia, let's go to the West Coast, where, in opposing strains, a centuries-old Haida song likens a woman's face to both flowers and snow.

> *Beautiful is she, this woman,*
> *as the mountain flower;*
> *but cold, cold, is she,*
> *like the snowbank*
> *behind which it blooms.*[91]

In *The Spell of The Yukon*, Robert Service captures the ambivalence inspired by a northern landscape where winter rules for most of the year. He calls it "the cussedest land that I know," but, in nearly the same breath, he testifies how "it twists you from foe to a friend." Part of the spell that turns fence-sitters like Service around has everything to do with snow.

"It's the beauty that thrills me with wonder / It's the stillness that fills me with peace. . . ." –Robert Service
John Poirier

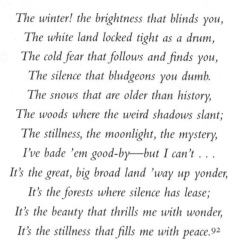

The winter! the brightness that blinds you,
The white land locked tight as a drum,
The cold fear that follows and finds you,
The silence that bludgeons you dumb.
The snows that are older than history,
The woods where the weird shadows slant;
The stillness, the moonlight, the mystery,
I've bade 'em good-by—but I can't . . .
It's the great, big broad land 'way up yonder,
It's the forests where silence has lease;
It's the beauty that thrills me with wonder,
It's the stillness that fills me with peace.[92]

In *River of the Angry Moon,* British Columbia writer Mark Hume reveals a common fascination with snow's peculiar power to cloak entire landscapes in a soft silent blanket that, in his words, "eats" all sound. It doesn't snow an awful lot in the Bella Coola region where he lives. Hume writes how farmers kept their trucks and cars "safely parked in driveways, covered with a 10-centimeter [4-inch] -thick blanket of fresh snow . . . knowing that sooner or later rain would come to wash the roads clear." Still, even in such country, snow works its soundless magic, eliciting—then eating—sounds that aren't even there.

> A heavy, wet snow has fallen during the night, spreading a humped white blanket across the estuary. Here and there clumps of tired blonde grass push through. A flock of mallards bursts from a tidal channel, scattering snow crystals in the air like pollen. Far out on the flats a drift of trumpeter swans stirs and shifts as an eagle circles overhead. Along the river, as the morning warms, clumps of snow drop from the tree branches, vanishing as they become water. The snow eats the sound of the river as it passes over its stone bed; it eats the sound of the forest. . . . I hear the Nuxalk drummers, but their song falls like snow into the river and becomes silent.[93]

Offer me one—only one—poem celebrating snow and I'll pick Robert Frost's "Stopping By Woods On A Snowy Evening." Here's a common man who, though spurred forward by the usual mundane obligations, or "promises," knows how to steal a sublime moment to watch a woods "fill up with snow" and attune his ears to "the sweep of easy wind and downy flake."

"He will not see me stopping here / To watch his woods fill up with snow. . . ."
—Robert Frost
John Poirier

Whose woods these are I think I know.
His house is in the village though;
He will not see me stopping here
To watch his woods fill up with snow.
My little horse must think it queer
To stop without a farmhouse near
Between the woods and frozen lake
The darkest evening of the year.
He gives his harness bells a shake
To ask if there is some mistake.
The only other sound's the sweep
Of easy wind and downy flake.
The woods are lovely, dark and deep.
But I have promises to keep,
And miles to go before I sleep,
And miles to go before I sleep.[94]

137

Not content with simply enjoying the shapes and sounds of snow as it falls, British sculptor Andy Goldsworthy made a pilgrimage to Canada's High Arctic to chisel the hardest snow on earth into sculptures resonating with the local landscape and culture. His book, *Touching North,* captures his fleeting masterpieces on film: rows of gothic arches that echo the mountainous backdrop, a white wall of fishlike arrows reflecting the Inuit dependence on wild game, a ring of four huge lock-stepped circles built at the North Pole, each facing different directions—all of them south. From March 22 to April 20, 1989, Goldsworthy, working with a bemused Inuk companion named Looty, supplemented what Emerson called "the frolic architecture of the snow" with fantastic shapes of his own design. His praises and ponderings of snow were transcribed from a small field tape recorder that often shook in his hands at temperatures approaching −40°C (−40°F).

Goldsworthy's aesthetic delight in snow roamed far beyond its sculpting potential to its sounds, its colors, and its importance to local people. At the back of his book is a glossary of 107 Inuktitut words for snow and ice—the longest such list I've seen in print. After just a month in the Inuit world, he humbly acknowledged that sorting out his insights about high arctic snow would be a life-long work in itself.

Friday 24th March

We stopped at some well wind-packed snow which Looty said was good for what I wanted but isn't what they would use for snow houses because this snow has too much air in it and he described it as "cold snow" . . . I found this snow is very strong and good to work with but it doesn't have a lot of tolerance. I am used to a snow that is frozen—this isn't frozen, it has been packed by the wind. I mean it is frozen but isn't frozen in the same way—it is not wet snow frozen. And it will suddenly just break if you handle it too roughly. The cold has given the snow here a particular quality and energy, and wet is not a part of its nature.

Sunday 2nd April

I came here with no tools but was willing to be guided by the Inuit. . . . The tools are very simple. But I have made one or two pieces of work that are just with my hands and I enjoyed that. It is important to me that the work that I make is about the snow and not about the edge of the saw and I must not be seduced by that edge.

Monday 3rd April

You walk on the snow and it makes a sound like I have never heard snow make before. It's almost a high-pitched, tensed-up sound and it's as if there is a lot of pent up energy in the snow. You tread on it and it is releasing that—it is like touching something and there is static, a shock. Each footstep is a bit like the shock.

"I have always considered snow and ice to be one of the most ephemeral of materials that I have ever worked with, but here it has a feeling of permanence." —Andy Goldsworthy

Though not hewn by Goldsworthy's hands, these sculptures on Yellowknife Bay were dubbed "Snowhenge," inspired perhaps by that paradoxical sense of snow's eternal strength.

Fran Hurcomb

Wednesday 8th April

Snow is stone—it is a white stone. Snow is like sand, ice is like slate—snow is sand and ice is slate. I have always considered snow and ice to be one of the most ephemeral of materials that I have ever worked with, but here it has a feeling of permanence and it makes me realize how rhythms, cycles and seasons in nature are working at different speeds in different places. The earth as a whole is probably in these cycles, going through different speeds and changing. Understanding those cycles is understanding the processes of nature.

Tuesday 18th April

I thought this could be an opportunity to work to bring out the blueness in the snow. The blueness appears when the snow is cut off from light—it is in the cracks and ridges, deep down in the snow. When you cut open a block and just prise it open there is this blue, blue light in there.

I am now at the North Pole

It is very difficult to say how I feel. I have worked very long hours and I haven't slept much in the last two days. The 24 day, it has been like time stopped still . . . I have a lot of feelings for here through what I have made, through touching the snow and finding good wind-blown snow. How I sort out those feelings is something that I'll do for the rest of my life.[95]

What Goldsworthy saw in snow were qualities like carving potential, tensile strength, or the ability to transmit sunlight when thinned. What his Inuit hosts saw was encyclopedic in comparison. Their traditional needs for food, shelter, locomotion, and navigation all hinged on an intimate knowledge of snow. Many contemporary Inuit retain this knowledge, drawing on it, for instance, to find the best kind of snow to make tea, or coffee, too, if that suits your tastes.

Coffee time during a caribou hunting trip near Baker Lake, Nunavut.
Tessa Macintosh

While writer and man-about-the-tundra David Pelly lived in Cambridge Bay, Nunavut, he once went out with some local seal hunters on a nippy midwinter morning when coffee time struck. In *Sacred Hunt,* he recounts how they zeroed in on the finest snow for the job.

> Knowing where to take the snow for meltwater at different times of the year is important. In this season, midwinter, the hunters collect snow to melt from the top layers of the generally flat snow cover on the sea ice, always being careful not to dig down near the surface of the salty sea ice. If there is an icy crust on top, they will take that, since it is denser than the snow beneath and is assuredly free of salt. Should they come across a snowdrift, where the wind has deposited an unusually high layer of snow, the granular snow found within the drift will have a higher water content than the fluffy wind-blown snow lying across the surface. In the spring, when the thaw begins, the sea ice freshens as the salt percolates downward, so ice taken from the top of an ice hummock will provide good fresh water even though it was originally frozen seawater. That good ice is easily identified by its color, which changes from the greenish turquoise hue of saltwater ice to a bluish white.[96]

From seeing, hearing, touching, and tasting snow (I've found nothing yet on its smell), California's cavalier naturalist John Muir takes us farther to a multisensory, whole-body appreciation of snow in "The Avalanche Ride."

> When the first heavy storms stopped work on the high mountains, I made haste down to my Yosemite den, not to "hole up" and sleep the white months away; I was out every day, and often all night, sleeping but little, studying the so-called wonders and common things ever on show, wading, climbing, sauntering among the blessed storms and calms, rejoicing in almost everything alike that I could see or hear: the glorious brightness of frosty mornings; the sunbeams pouring over the white domes and crags into the groves and waterfalls . . . the great forests and mountains in their deep noon sleep; the good-night alpenglow; the stars; the solemn gazing moon, drawing the huge domes and headlands one by one glowing white out of the shadows hushed

and breathless like an audience in awful enthusiasm, while the meadows at their feet sparkle with frost-stars like the sky; the sublime darkness of storm-nights, when all the lights are out; the clouds in whose depths the frail snow-flowers grow. . . .

The fertile clouds, drooping and condensing in brooding silence, seem to be thoughtfully examining the forests and streams with reference to the work that lies before them. At length, all their plans perfected, tufted flakes and single starry crystals come in sight, solemnly swirling and glinting to their blessed appointed places; and soon the busy throng fills the sky and makes darkness like night. . . .

After snow-storms come avalanches, varying greatly in form, size, behaviour and in the songs they sing. . . . Most delightful it is to stand in the middle of Yosemite on still clear mornings after snow-storms and watch the throng of avalanches as they come down, rejoicing, to their places, whispering, thrilling like birds, or booming and roaring like thunder.[97]

Eager to see as many avalanches as possible, Muir set out "one fine Yosemite morning after a heavy snowfall." His goal was to climb a side canyon to the top of a commanding ridge in order to gain a view of the forest and peaks "in their new white robes." But his heavy trampling near the canyon head triggered an avalanche that swished him down to the foot of the mountain "as if by enchantment."

The wallowing ascent had taken nearly all day, the descent only about a minute. When the avalanche started I threw myself on my back and spread my arms to try to keep from sinking . . . though I was tossed here and there and lurched from side to side. When the avalanche swedged and came to rest I found myself on top of the crumpled pile without a bruise or scar. This was a fine experience . . . This flight in what might be called a milky way of snow-stars was the most spiritual and exhilarating of all the modes of motion I have ever experienced. Elijah's flight in a chariot of fire could hardly have been more gloriously exciting.

After a ride like that—which, by the way, I cannot recommend— how else might you appreciate snow just being snow? Yukon poet Erling

Friis-Baastad gives us another, safer option: Go out snow-watching some clear frigid night. In "Spending Your Death In The Yukon," he invites us to join the company of ghosts set aglow by the northern lights and to revel in the spellbinding brightness of moonlight on snow.

> *If the winter nights up here*
> *are perfect for ghost watching*
> *they are even more perfect*
> *for being a ghost.*
> *Imagine a soul—by day*
> *it felt unclaimed and gray—*
> *suddenly set aglow*
> *by a bombardment of northern lights.*
> *Then, picture that scene framed*
> *by the leafless willows, encrusted*
> *with hoarfrost.*
> *They make good company,*
> *like the ghosts of trees.*
> *And the moon! The moon over all*
> *this snow is so bright*
> *even a shade would cast a shadow.*[98]

Robert Frost was another admirer of moonlit snow, though by the time he wrote "Evening in a Sugar Orchard," there were only a few patches left, reminding him of polar bears asleep on the dark floor of his maple woods.

> *The moon, though slight, was moon enough to show*
> *On every tree a bucket with a lid,*
> *And on black ground a bear-skin rug of snow.*[99]

At Home in the Snow

In *Good Hours* Robert Frost pays tribute to that singular comfort that comes from being tucked in a warm home while deep snow lays round about. His view is from the outside in, vicariously enjoying the companionship within the cottages he passes as he saunters to the edge of the village and beyond. But a palpable coziness comes blazing through those cottage windows that are "up to their shining eyes in snow." This feeling is magnified on the poet's return journey when those same windows are blackened as the village sleeps.

I had for my winter evening walk—
No one at all with whom to talk,
But I had the cottages in a row
Up to their shining eyes in snow.
And I thought I had the folk within:
I had the sound of violin;
I had a glimpse through curtain laces
Of youthful forms and youthful faces.
I had such company outward bound.
I went till there were no cottages found.
I turned and repented, but coming back
I saw no window but that was black.
Over the snow my creaking feet
Disturbed the slumbering village street
Like profanation, by you leave,
At ten o'clock of a winter eve.[100]

In *Songs of the Dream People*, James Houston captures the same feeling, experienced this time from the inside out. Who could be more familiar with that sense of snugness in the snow than a Greenland Inuk waiting out a storm in an igloo?

It is still in the house;
The snowstorm wails outside.
My little boy is sleeping on my back;
His stomach is bulging round.
Is it strange if I start to cry with joy?[101]

As Reverend Thomas Sims stepped up to the pulpit of Toronto's Bond Street Congregational Church on Sunday morning, January 26, 1896, he faced a rather disgruntled congregation. After a weeklong snow-storm, his flock was thoroughly fed up with snow and all the troubles and pain it had wreaked. To cool tempers and warm hearts Sims spoke not of God's wrathful power and might—in other words, "You asked for this, folks"—but of the hidden blessings showered upon them by the storm. For almost an hour he held forth on "The Treasures Of Snow," among which was snow's paradoxical role in protecting life, human or otherwise.

Beneath its fleecy folds, God tucks in flower and grass, herb and root of tree as snugly as a mother cradles her cherished babe. How green and thrifty a field of fall wheat emerges after a long, hard winter, when the snow has lain steadily upon it! . . . In Ontario's early days, when the land was thickly wooded, and the snow lay still and late, potatoes were often left in the ground all winter, to be dug up sound and good in the spring . . . Ensconced in his snow hut, the Esquimaux passes his Arctic winter in comfort, the only times when the cold seriously troubles him being when the weather has begun to grow mild and the warmth of his snow dwelling makes it damp. Lost in the fierce blizzards of the northern prairie, many a settler has saved his life by creeping under his upturned sleigh, and allowing the snow to drift over him. So does God ordain that snow, a product of the cold, shall be to herb and beast and man a protection from the cold that would otherwise destroy them.[102]

This eight-person igloo was built at a barrenland mining exploration camp. Rather than snowslides, the main threat to this structure was propeller wash from airplanes roaring past it with diamond-bearing ore. "It stood up well," says the proud builder and photographer.

Lorne Schollar

"The Father's Song" is a traditional Inuit song in which a shaman
addresses the spirit of a snowslide that threatens their igloo home. The slide
hasn't happened, may never happen, or might have been bound to happen,
were it not for the shaman's stirring appeal.

> Great snowslide,
> Stay away from my igloo,
> I have my four children and my wife;
> They can never enrich you.
> Strong snowslide,
> Roll past my weak house.
> There sleep my dear ones in the world.
> Snowslide, let their night be calm.
> Sinister snowslide,
> I just built an igloo here, sheltered from the wind.
> It is my fault if it is put wrong.
> Snowslide, hear me from your mountain.
> Greedy snowslide,
> There is enough to smash and smother.
> Fall down over the ice,
> Bury stones and cliffs and rocks.
> Snowslide, I own so little in the world.
> Keep away from my igloo, stop not our travels.
> Nothing will you gain by our horror and death,
> Mighty snowslide, mighty snowslide.
> Little snowslide,
> Four children and my wife are my whole world,
> all I owe,
> All I can lose, nothing you can gain.
> Snowslide, save my house, stay on your summit. [103]

Inner Tracks and Trails

It is the rare poet or writer out there who has trained the inner eye on snow,
seeing it not as fierce, friendly, or even fluffy, but as something beyond all
those overdone labels, something pointing to the transcendent. A master
wordsmith inclined to look deeply into snow's nameless face can dissolve the
boundary between self and snow—if only for a poetic moment or two. In

"The Snow Man," Wallace Stevens invites you to cross that boundary by adopting "the mind of winter," suspending all judgment of its winds, and to simply listen in the snow. In the deft hands of this poet, snow becomes the springboard to our identity with all nature. As Stevens himself once remarked, this poem is about "the necessity of identifying oneself with reality in order to understand and enjoy it."

> *One must have a mind of winter*
> *To regard the frost and the boughs*
> *Of the pine-trees crusted with snow;*
> *And have been cold a long time*
> *To behold the junipers shagged with ice,*
> *The spruces rough in the distant glitter*
> *Of the January sun; and not to think*
> *Of any misery in the sound of the wind,*
> *In the sound of a few leaves,*
> *Which is the sound of the land*
> *Full of the same wind*
> *That is blowing in the same bare place*
> *For the listener, who listens in the snow,*
> *And, nothing himself, beholds*
> *Nothing that is not there and the nothing that is.*[104]

By whatever label—deep ecology, earth-centered spirituality, or neo-paganism—this notion of finding our true identity in nature is as old as the hills and as fresh as the snow. In *The Stars, The Snow, The Fire,* John Haines writes of his hermit's journey through the northern wilderness, where trails into the hills and across the snow lead as much inward as outward.

> The physical domain of the country had its counterpart in me. The trails I made led outward into the hills and swamps, but they led inward also. And from the study of things underfoot, and from reading and thinking, came a kind of exploration, myself and the land. In time the two became one in my mind. With the gathering force of an essential thing realizing itself out of early ground, I faced in myself a passionate and tenacious longing— to put away thought forever, and all the trouble it brings, all but the nearest desire, direct and searching. To take to the trail and

not look back. Whether on foot, on snowshoes or by sled . . . a runner track in the snow would show where I had gone. Let the rest of mankind find me if they could.[105]

"Whether on foot, on snowshoes or by sled . . . Let the rest of mankind find me if they could." —John Haines
Northern News Service

Whether those inner tracks across the snow tell of a solo passage or a caravan of cotravelers, they lead to the same destination: an inner place of wholeness, freedom, and connectedness—the time-tested fruits of a spiritual journey. In *Earth Wisdom,* Dolores LaChapelle captures a timeless moment when these fruits are shared by a company of skiers finding the perfect "way" down a mountain. Surrendering themselves to gravity, they fly as one body over the dips and contours of deep powder snow.

> The freedom, grace, and joy of this togetherness in the powder "world" occurs in response to the gift of the sky: unbroken snow. This is most easily skied in direct response to the earth's gravity—down the fall line—but the dips and contours of the earth automatically lay out the "way" to follow; and for the skilful skiers, there is only one best "way" for each, so all can ski together at top speed and still flow with one another and with the earth . . . Just as in a flight of birds turning through the air, no *one* is the leader and none are the followers, yet all are together; so also the powder snow skiers are all together effortlessly, because they are appropriating, responsively conforming themselves to the earth and sky in their "world," thus there are no collisions. Each human being is free to follow his own path.[106]

In *Rolling Thunder,* anthropologist Doug Boyd chronicles the life and beliefs of a modern-day shaman from the American Midwest. As spiritual leader for the Cherokee and Shoshonee tribes, Rolling Thunder guards a wealth of mysterious knowledge that includes the power to cure disease, perform exorcisms, levitate objects, even bring down rain or snow. The source of Rolling Thunder's powers springs from his intimate relationship with nature, to the point where, for instance, snow and shaman are one.

> For him there are no weeds, no mosquito bites, no unwanted rains. There are no dangerous plants or animals. For him there is no fear. The wind and the rain, the mosquitoes and the snakes are all within him. His consciousness extends to include them within its very being. What the life of Rolling Thunder communicates is that when someone identifies himself, not with his self-image or his thinking process but with the flowers, the snow,

and all manifestations of the life force, he can do the things of which Rolling Thunder speaks.[107]

As much as Boyd may root for your personal development, you may have no interest in making pigs fly or calling down the snow. Instead you might fix your spiritual sights on more modest goals, like becoming an all-round nicer person. To meet this formidable challenge, Reverend Thomas Sims, in *Treasures of Snow*, suggests we look for inspiration in snow's beauty. Though his turn of phrase may convey more than a century's worth of dust, his message carries perennial value: the more beauty we see in the outer world of nature, the more we will find within (The reverse of course is also true.). This discovery, says Sims, will in turn be reflected in those tremendous trifles of our daily life.

> The beauty of the snow! Is it not a gleam of the beauty of God?
> Is it not intended to suggest what he would have all his children
> be? Holiness is more than aesthetics, but holiness and beauty are
> indissolubly married. The law of beauty is a law of God, a law
> not simply of his will, but of his nature. . . . Consider the beauty
> of the snow, today, and try to realize that there is in it an expres-
> sion of God's will concerning you. And if, in harmonizing the
> outward and the inward, you put another touch of beauty and
> grace on your homes, on your speech, on your manners, and on
> your charities, after considering the treasures of the snow, many
> of you will be doing the Lord's will in that.[108]

During a rollicking New England snowstorm in 1838, Ralph Waldo Emerson found he had little presence of mind for preachers, however well-intentioned. What the snow could teach him held far more sway. Always the fearless philosopher, he later shared this experience to an audience of would-be preachers in his "Address to the Harvard Divinity School." Keep in mind that Emerson's first career was as a preacher for the Unitarian church, a modern spokesman of which proclaimed ironically that his Harvard address remains "perhaps the best known sermon ever preached in America."[109]

> I once heard a preacher who sorely tempted me to say I would
> go to church no more. Men go, thought I, where they are wont
> to go, else had no soul entered the temple in the afternoon. A
> snow-storm was falling around us. The snow-storm was real, the

preacher merely spectral, and the eye felt the sad contrast in looking at him, and then out of the window behind him into the beautiful meteor of the snow.[110]

"Consider the beauty of the snow. . . ."—Thomas Sims
John Poirier

Whether viewed from a church window or from a moving dogsled, the freshness of snow, if nothing else, may prod us to clean up our outer and inner acts. Who could deny that the simple gesture of mouthing a fistful of virgin snow has a purging effect on the soul? In *Enduring Dreams: An Exploration of Arctic Landscape,* John Moss asks you to "remember fresh snow melting in your mouth when you were a child. . . . That was what clean tasted like; the word and the sensation, inseparable."[III] What happens when we turn our backs on snow and deny its purifying grace? Michael Yates, a poet and playwright from Squamish, British Columbia, explores this theme in "The Hunter Who Loses His Human Scent." This haunting tale follows the journey of a man who "has been moving north always . . . On his back he carries, pack, snowshoes, and rifle; apparently he has carried them always . . . presumably since his birth . . ." Near the journey's end, just "a short distance" from his true northern home, the dogsledding pilgrim makes one fatal mistake, forgetting the rite of melting snow in his mouth. He dies alone in pleasant contemplation of an enveloping blizzard.

"A man, warmly dressed, in perfect health, mushing his dogs a short distance between two villages, never arrives . . ." –Michael Yates
Tessa Macintosh

A man, warmly dressed, in perfect health, mushing his dogs a short distance between two villages, never arrives. He has forgotten to reach down, catch a little snow in his mitten, and allow it to melt in his mouth.

For a reason neither he nor his dogs understand, he steps from the runners of his sled, wanders dreamily—perhaps warmly, pleasantly—through the wide winter, then sits to contemplate his vision, then sleeps.

The dogs tow an empty sled on to the place at one of the two villages where they're usually fed.

While those who find the frozen man suspect the circumstances of his death, always they marvel that one so close to bed, warmth, food, perhaps family, could stray so easily into danger.[112]

Browsing through the literature on snow, fans of psychologist Carl Jung might say that there was something archetypical about the purging power of melting snow in your mouth. It seems to be a deep-seated motif that jumps from the consciousness of many cultures. In *Sacred Hunt,* David Pelly describes how the Inuit of northern Canada carry this sacred rite into the art of seal hunting.

As a mark of respect, many Inuit groups had a tradition of giving the dead seal a "drink" of fresh water before the butchering began. How this was done varied. In the northwest corner of Hudson Bay, 77-year-old Mikitok remembers seeing and following this practice: "When we got the *nattiq* back to the iglu, never out where it was caught, we took some snow, dipped it in melted water, and put it in the seal's mouth." In the central Arctic, another respected elder, Analok, described it like this: "Before you cut up a seal, you would get some meltwater from your mouth and pour it into the seal's snout. First you melt fresh snow in your mouth, then you pour it in the seal's mouth." In both regions, the hunter's wife performed the ritual . . . Some Inuit believed that a seal refreshed in this manner would be more likely to return again in the form of another seal for another drink. In all cases, the idea was to appease the seal's spirit. "That custom was passed on from generation to generation," said Analok, "so that the hunter would catch seals again in the near future, so that the seal would be renewed."[113]

Traveling through the same country in 1922, anthropologist Knud Rasmussen observed how the Inuit drew on snow's protective power to avoid upsetting any higher authorities.

When a seal has been cut up and lies in pieces on the floor, a lump of fresh snow is laid on the spot where its head was, and trodden down there. The Sea Spirit does not like women to tread on the spot where the seal's head has lain.[114]

In "A Winter Eden," Robert Frost declares that a good day in a snow-filled forest—even a good *hour*—is "as near a paradise as it can be." Snow lifts us heavenward even as it boosts a waking bear to new heights of scratching and feasting on the bark of a wild apple tree. Though Frost paints a gorgeous picture of outer beauty, paradise is a state of mind and heart which he clearly was close to when he viewed this scene.

> *A winter garden in an alder swamp,*
> *Where conies now come out to sun and romp,*
> *As near a paradise as it can be*
> *And not melt snow or start a dormant tree.*
> *It lifts existence on a plane of snow*
> *One level higher than the earth below,*
> *One level nearer heaven overhead,*
> *And last year's berries shining scarlet red.*
> *It lifts a gaunt luxuriating beast*
> *Where he can stretch and hold his highest feast*
> *On some wild apple tree's young tender bark,*
> *What well may prove the year's high girdle mark.*
> *So near to paradise all pairing ends:*
> *Here loveless birds now flock as winter friends,*
> *Content with bud-inspecting. They presume*
> *To say which buds are leaf and which are bloom.*
> *A feather-hammer gives a double knock.*
> *This Eden day is done at two o'clock.*
> *An hour of winter day might seem too short*
> *To make it worth life's while to wake and sport.*[115]

Melting into Spring

Snow comes, snow goes. It pounds on your door; it tickles your nose. Rummaging through a treasure trove of word pictures about snow, I found a few gems that pay special tribute to its springtime passage. For instance, New Hampshire poet Jane Kenyon, in *The Boat of Quiet Hours,* features this fragile time of year in her chronicle of inner and outer storms. One admirer describes her poetry as being "permeated with the steadying influence of nature." Here the poet, "greedy for unhappiness," stumbles upon a warm boulder set amid melting snow. She sits a spell, just sits, then rises "chastened and calmed."

> *There comes a little space between the south*
> *side of a boulder*
> *and the snow that fills the woods around it.*
> *Sun heats the stone, reveals*
> *a crescent of bare ground: brown ferns,*
> *and tufts of needles like red hair,*
> *acorns, a patch of moss, bright green . . .*
> *I sank with every step up to my knees,*
> *throwing myself forward with a violence*
> *of effort, greedy for unhappiness—*
> *until by accident I found the stone,*
> *with its secret porch of heat and light,*
> *where something small could luxuriate, then*
> *turned back down my path, chastened and calmed.*[116]

Late spring snow dusts the Northwest Territories' Trout River, swollen with melt-water.

John Poirier

In the melting snows of his Kentucky home, farmer-philosopher Wendell Berry hears the sound of birdsong as he welcomes a long-awaited "return to earth."

> *Through the weeks of deep snow*
> *we walked above the ground*
> *on fallen sky, as though we did*
> *not come of root and leaf, as though*
> *we had only air and weather*
> *for our difficult home.*
> *But now*
> *as March warms, and the rivulets*
> *run like birdsong on the slopes,*
> *and the branches of light sing in the hills,*
> *slowly we return to earth.* [117]

When the snow starts to melt in earnest, spring fever can hit the poetically inclined pretty hard. Erling Friis-Baastad harnessed some of that restlessness in *Yukon Spring,* admitting that, while "lost upon a slim trail" this time of year, he may feel like "an anonymous composer of hymns" with "the latest plague on his heels."

> *Now that the snow has been beaten*
> *back into small dour patches*
> *between the black spruce,*
> *I cannot sit still.*
> *Each morning finds me*
> *out hiking beneath*
> *the new sun,*
> *striding purposefully*
> *as if I had somewhere*
> *to go.*
> *There seems to be a small*
> *inarticulate religious fanatic*
> *inside me. I must bear him*
> *with me whenever I go. . . .* [118]

As the rains of spring wash the last shards of snow away, I am often gripped by a pronounced ambivalence—nostalgic for the winter that was and eager for the looming spring. Underneath this turbulence is my reluctant assent to life's evanescence, reflected in the mirror of fading snow. In "A Patch of Old Snow," Robert Frost plays with these emotions, sparked by a lonely heap of dirty snow.

> *There's a patch of old snow in a corner*
> *That I should have guessed*
> *Was a blow-away paper the rain*
> *Had brought to rest.*
> *It is speckled with grime as if*
> *Small print overspread it,*
> *The news of a day I've forgotten—*
> *If I ever read it.*[119]

Have you guessed by now that I like Robert Frost? Here I give him a well-deserved last word on snow's passage into oblivion. In "The Onset," Frost celebrates the triumph of returning life over "winter death" and reveals his fascination with the way snow disappears, leaving nothing white except "here a birch / And there a clump of houses with a church."

> *I know that winter death has never tried*
> *The earth but it has failed: the snow may heap*
> *In long storms an undrifted four feet deep*
> *As measured against maple, birch and oak,*
> *It cannot check the peeper's silver croak;*
> *And I shall see the snow all go down hill*
> *In water of a slender April rill*
> *That flashes tail through last year's withered brake*
> *And dead weeds, like a disappearing snake.*
> *Nothing will be left white but here a birch,*
> *And there a clump of houses with a church.*[120]

Chapter 5
Confronting Snow

Snow is the bleak reality of a stalled car spinning wheels impinging on the neat time schedule of our self importance.

                             –Farley Mowat, *The Snow Walker*[121]

Snowed Under by Progress

"The growth of cities has transformed one of nature's glorious wonders into a perplexing and costly problem," writes Blake McKelvey in *Snow in the Cities*.[122]

In North America, the trouble all started when early Europeans began setting up shop here, gearing up for winter as if embarking on a picnic in the park. Prolonged subzero temperatures, blizzards, and backbreaking snowfalls apparently came as an unpleasant surprise to many early settlers whose ridiculously ill-suited clothing and architecture were no match for this new brand of winter. For instance, in the tiny French settlement of Tadoussac, Quebec, where the Saguenay River enters the St. Lawrence, only eleven of thirty people survived the winter of 1600. Down in Sagadahoc, Maine, a group of British settlers found the winter storms of 1608 a little too "vehement" for their liking. The very next spring all survivors beat it straight for England without looking back.

Eventually the more pliant Europeans did catch on, picking up survival tips from aboriginals who were as well adapted to New World winters as the snowshoe hare or the caribou. Cold, drafty stone buildings gave way to more substantial wood structures that were banked high with snow all winter for extra insulation. Beaver hats, wolf-trimmed clothing, thick moosehide footwear, and a sturdy pair of snowshoes became standard equipment for early fur traders and voyageurs busy pushing back the continent's economic frontiers. With improved sled designs and a team of willing horses, snow and ice became a boon to hauling heavy gear and supplies between scattered settlements.

Winter had a winnowing effect on early North American settlers. Sturdy, well-chinked log structures banked with snow increased the odds of their seeing the next spring.

Lorne Schollar

And when the big snowstorms hit, early settlers simply holed up for a few days, secure with their homegrown provisions, crafts, and amusements. This was still an era when "the loss of a week or two of time was of very little consequence."[123] But as pioneer settlements grew into towns, snow threw new curves at those settling into urban life.

Boston was already a town of twelve hundred souls when the first big bad storm hit in February 1642. It was, in fact, the worst on record, if you count what one diary called "the memory of friendly Indians." It got so cold that Boston Harbor froze over entirely in a matter of days. At the onset of the storm, several men drowned struggling to bring boats into port. By the time it blew over, men were driving large teams of horses and carts loaded high with precious firewood over ice 15 centimeters (6 inches) thick. During the next few weeks, more deaths and frozen limbs followed in the wake of uncertain ice conditions, dwindling firewood, and swamped vessels carrying critical supplies to this swelling mass of humanity in the middle of nowhere. Apparently most survivors of this "landmark winter" owed their good fortune to nothing less than "God's mercy." (The fact that many of the surviving journals from that era were written by men of the cloth may have contributed to this impressive dose of divine grace.)

In those days, a highly accurate gauge for measuring the severity of snowstorms was the number of Sundays the church doors remained locked and bolted. In March 1698, after a particularly thorough drubbing of Cambridge, Massachusetts, where snow lay "three feet and a half deep on a level," things got so desperate that, once regular church services resumed, the preacher invited his parishioners to fortify their prayers for mercy with a "public fast" to petition God for relief from the snow.[124]

While the buffetings of snowstorms often brought early settlements to their knees, small farms in rural districts were much more resilient.
John Poirier

Boston and other New England towns bore the brunt of the Great Storm of 1717. After a volley of bracing snow squalls, Boston reported more than a meter (42 inches) of snow in its streets, while high winds had curled drifts up against many second-storey windows. Not a soul was seen from rural districts until a lone mail carrier arrived on snowshoes two weeks later, reporting what most of the town already knew—all overland supply routes from the interior were chock-a-block with snow. After marveling at God's tempestuous handiwork, Reverend Cotton Mather described in his diary the coming of another blockbuster a few days later. "Another snow came on which almost buried ye memory of ye former, with a Storm so famous that Heaven laid an Interdict on ye Religious Assemblies throughout ye Country." On that note, he turned the lock on his own church doors, which remained sealed by drifts for the next two Sundays. When he finally opened for business, he bid his congregation to mark "a day of humiliations and supplication," sprinkled with "suitable acts of charity" to help fend off further retribution from on high.[125]

Throughout the burgeoning days of colonial settlement, the tool of choice—and often the only tool—for dulling the toughest blows of winter was prayer. Sermons of the day stoked the prayers of the faithful with fuel sometimes hot enough to melt a small mountain of snow. Take this excerpt from a sermon delivered at Narragansett, near Providence, in March 1741 by one Reverend James MacSparren. For full effect, read this out loud with a fiery Scottish accent.

> The recent suffocating snows are a warning of God's Vengeance
> on us for our Ingratitude to his Goodness and our Transgression
> of His Law. . . . Would we therefore be relieved of the Burden
> and Inconvenience of the Winter . . . we must propitiate the God
> who alone can invite us to sing and say in the Language of
> Solomon's Song, "Lo the Winter is past. The Flowers appear on
> Earth, the time of the singing of Birds is come, and the Voice of
> the Turtle is heard in the Land."[126]

We can only assume that the prayers of an unnamed clerk who kept the records for the First Church in Cambridge, Massachusetts, must have been heard during the next winter of heavy snows in 1747-48. After jotting down a few lines about the thirty-odd snows that had dumped 1.5 meters (5 feet) of snow and lasted well into March, he cheerily concluded that "on the twelfth day of April, I had my garden sowed and planted with onions, carrots, parsnips and beans."[127]

By 1780 a mounting congestion of horse-drawn vehicles had added new challenges to winter life in urbanizing North America. According to Blake McKelvey, "a new contingent of weather-watchers had also arrived; equipped with thermometers, barometers, and clocks, they were more concerned with collecting specific readings and comparing them with past records and with neighbouring reports than with speculating on a Heavenly message."[128] Traditional attitudes linking severe snowstorms to degenerate human behavior gave way to more pragmatic questions, like, "How the hell are we going to get rid of all this snow?"

Over the next few decades, as towns grew into cities, so did snow's potential to throw a wrench in the urban machine. Fast-forwarding to January 3, 1856, we find the diary entry of George Templeton Strong, who left us a revealing snapshot of the bedlam created when a thumping good storm hit New York City.

> This is a stern winter. Saturday's snowstorm was the severest for many years past. The streets are like Jordan, "hard roads to travel." One has to walk warily over the slippery sidewalks and to plunge madly over crossings ankle-deep in snow, in order to get uptown and down, for the city railroads are still impracticable and walking (with all its discomforts) is not so bad as the great crowded sleigh-caravans that have taken the place of the omnibi. These insane vehicles carry each its hundred sufferers, of whom about half have to stand in the wet straw with their feet freezing and noses tingling in the bitter wind, their hats always on the point of being blown off. When the chariot stops, they tumble forward, and when it starts again, they tumble backward, and when they arrive at the end of their ride, they commonly land up to their knees in a snowdrift, through which they flounder as best they may, to escape the little fast-trotting vehicles that are coming straight at them. Many of the cross streets are still untraveled by anything on wheels or runners, but in Broadway, the Bowery, and other great thoroughfares, there is an orgasm of locomotion. It's more than a carnival; it's a wintry dionysiaca.[129]

(To save you the trouble of looking it up, as I had to, a "dionysiaca" is a sensual, wild, unrestrained party; in this case, an overindulgence in snow.)

The word *blizzard* came into vogue around 1870, after a series of sock-blowing storms pummeled the American Midwest. In St. Paul, for instance, the 1865 New Year's festivities ground to an icy halt when the thermometer suddenly plunged to −35°C (−31°F). Reports of snowdrifts up to 3 meters (10 feet) high boosted the region's newspaper sales over the next few winters.

Bring together a blizzard of continental proportions and North America's largest, cockiest city, and you've got a recipe for winter drama the likes of which have never been seen before or, some say, since. Thus was the fate of New York City during the tempest of March 1888. It began without fanfare on March 6 as a modest storm center over the North Atlantic drifted lazily westward. Swirling counterclockwise, its southern arm began sucking a moisture-rich air mass over the Carolina coast, eventually driving it northward up the eastern seaboard. Meanwhile the storm center's northern arm caught hold of a frigid arctic air mass drifting over central Canada. By the afternoon of March 11, the inevitable collision of warm moist air with the dry arctic stream occurred somewhere over the heads of New Yorkers, heralded by high winds and sopping rains. As night fell the beast transformed into a blinding snowstorm.

Over two million citizens woke dazed and blinking at the sight of streets meringued with 1.5-meter (5-foot) drifts. By nightfall of March 12, little moved on New York's streets except the wind, now gusting to 120 kilometers (70 miles) per hour. The second biggest city on earth was "as quiet as a tomb." Four-fifths of its ten thousand telephones were silent. Most of its electric lights and gas lamps were blacked out. What is really quite amazing, however, is that somewhere indoors, the city's great newspaper presses rolled throughout the storm. On March 13, right on schedule, the *New York Herald* hit the streets—such as they were.

"The only lights to be seen up and down the dreary waste of snow," wrote one intrepid reporter, who seems to have partied the night away, "were those that shone from the comfortable interiors of several saloons that are scattered along Broadway, and it is needless to say that they were well filled with blizzard-filled people."[130]

While the saloon and hotel business "reaped a harvest," this one storm delivered a pivotal kick in the face to urban America's belief in its conquest of nature. "In looking back at the events of yesterday," went another story in the "Blizzard Edition" of the *New York Times,* "the most amazing thing to the residents of this great city must be the ease with which the elements were able to overcome the boasted triumph of civilization, particularly . . . our superior means of intercommunication."[131]

When ice blocked the rivers, even Manhattan's Brooklyn Bridge, the engineering marvel of its day, was shut down. A commentary from the *Evening Sun* vividly shows how the Big Apple became a plaything in the hands of the Great Blizzard of '88.

It was as if New York had been a burning candle upon which nature had clapped a snuffer leaving nothing of the city's activities but a struggling ember . . . The streets were littered with blown-down signs, tops of fancy lamps, and all the wreck and debris of projections, ornaments and movables. Everywhere horse cars were lying on their sides, entrenched in deep snow, lying across the tracks, jammed together and in every conceivable position. The city's surface was like a wreck-strewn battlefield.[132]

Some of the blown-down signs were replaced by therapeutic stand-ins, stuck here and there in the snowdrifts probably by well-meaning children who had nothing better to do. Their messages read: "Keep off the Grass," "Do Not Pick the Flowers," and "Important Notice: This is Twenty-Third Street." My hope is that those youngsters took care in erecting their signs since, with the indefinite postponement of all funerals, the corpses of loved ones had to be stored in snowbanks for safekeeping.

It took three days for the Great Blizzard of '88 to finally blow itself out, leaving at least four hundred dead in New York city alone. Several days went by before New Yorkers learned—via underwater cable from London—that Boston had been hit equally hard. So vast was the blizzard's footprint that out on the American plains and Canadian prairies, over a hundred more people died, along with thousands of cattle.

In the meantime, Snowflake Man Wilson Bentley was having the time of his life over in Jericho, Vermont. During the most disastrous snowstorm in North American history, he was out in the barn photographing snow crystals like never before. During one of the storm's wildest tantrums, he took a dozen good snaps, the most he had ever taken in a single day. "All splendid crystals," he scribbled in his notebook. "Taken in great blizzard. Cloudy. Snowing hard." Well pleased with the results, he later wrote to an interested meteorologist with the hopes of selling a few images: "I have secured sets . . . from almost all the great snow storms and blizzards from March 12, 1888 to March of this year . . . It will give me great pleasure to receive an order from you, for a collection of these gems, wrought by the blizzards."[133]

An editorial in the Boston *Daily Globe,* written many years later, spoke to the "desperate vulnerability" of modern civilization when besieged by snow—epitomized by the Great Blizzard of '88. In a perverse sort of way, the editor goes on to proclaim that a good snowstorm can unleash latent reserves of inebriating valor and spunk as urbanites buck up to its many trials. "Does it take a blizzard to jolt us out of the drab monotony of our mechanized routines and give us a taste of that powerful intoxicant, man's hand-to-hand conflict with nature, which has made his life on this earth a thing of vivid interest? Then give us the blizzard."[134]

The past century presents a rip-roaring litany of calamities created by urban snow, spiced with a healthy mix of ennobling charity, and, more often than not, redeeming good humor.

February 1, 1916. Victoria, British Columbia. On the coattails of three straight weeks of below freezing weather and knee-high snow, the capital of Canada's banana belt succumbs to a snowstorm that would come as no surprise in a place like Tuktoyaktuk. But in a city known as Gardener's Heaven, where snow shovels sit idle for years at a stretch, street-plugging drifts one-storey high are something of an anomaly, to say the least. Police and firefighters flail away at the snow with the city's measly arsenal of a hundred shovels. Clearly more steam is needed for the job. Up jump members of "the fair sex," wrote the *Daily Province,* who shed their customary long skirts and pointy shoes for men's trousers and boots, apparently causing much "titillation" in the streets. Eventually they are digging shoulder to shoulder with a thousand Canadian soldiers who, after two weeks of hard labor, help Victoria find its feet again.

March 26, 1930. Chicago, Illinois. A low-pressure cell blasts down from Lake Superior and disgorges enough snow on America's motor city to keep an army of forty thousand shovelers busy for many days. Some city engineers, perhaps anxious to get behind the wheel of a fancy new snowplow, feel that sophisticated Chicago has moved beyond the age of shovelry. This is a hard sell during the Depression, when one pass of a plow could scoop work out from under a thousand hungry diggers. Besides, one humorist observes, "the snow shovel will find the boundary line between two lots more accurately than the best surveyor."[135] A *New Yorker* cartoon from this era shows a zealous shoveler, elated to be employed, removing not only the snow but the pavement, the soil, plus the gas and sewer lines below. Standing over him with arms waving is his distraught supervisor. "No, no, McNamara," he shouts. "Just that white fluffy stuff on top!"[136]

February 14, 1940. Boston, Massachusetts. The city is hit with a Valentine's Day blizzard. Sixteen thousand men and fifteen hundred trucks, plows, blowers, and graders are pressed into service for five full days. City officials later conclude that by adopting a tighter, more military approach to snow removal they could triumph over future blizzards with safety and ease. True to these warring times, the *Daily Globe* reflected similar tones by describing the storm as the "Snow Blitzkrieg of 1940," adding that "deep snow pushed the clock back" to more old-fashioned winters. Later that year, on Armistice Day no less, another blizzard swallows the Midwest, killing nearly 150 people. So much for better planning.

December 11, 1944. Toronto, Ontario. With soldiers still falling overseas, a modest afternoon snowfall helps Torontonians muster a Christmas spirit. That evening, the snowfall tilts about 45 degrees as it quadruples in volume and velocity. Sixty centimeters (2 feet) of snow falls through the night with more on the way. The next morning, downtown pedestrians are knee-deep in snow with occasional drifts at their shoulders. Mayor Conboy takes to the radio, urging all but war workers and essential personnel to kindly go straight home. Thousands of determined (or deaf?) workers ignore his message and are eventually forced to abandon their cars willy-nilly in the streets. Pregnant women in labor arrive at St. Michael's Hospital via coal trucks, delivery vans, army jeeps, a cement mixer, and on foot. One Margaret Haxton, scheduled to be married that day, trades her wedding gown for a ski suit, schusses over a 1.5 kilometers (1 mile) to St. John's Catholic Church, and arrives at the altar just in time, with snow trickling down her beaming face. That night the Toronto Symphony displays equal grit by honoring its engagement at Massey Hall. Amazingly there is barely an empty seat in the house. Near the back of the hall sits a young nurse in training—my mother. She remembers conductor Sir Ernest MacMillan profusely thanking the audience for showing up on a wild night like this. A black-tie occasion as always, many had gotten there by skiing down the middle of Yonge Street.

December 27, 1947. New York City. While Boxing Day parties wind down, 60 centimeters (2 feet) of snow falls through the night. The next day's edition of the *New York Times* concludes that nature's grip on the city is as tight as ever. "Our city has greatly changed in those sixty years since the Great Blizzard, but man is still at the mercy of the elements as anyone who looked out yesterday on New York's streets could see. . . . Life [became] disordered then frenzied as distressed workers struggle to get home."[137] While the Big Apple is clutched once again in "a great paralysis," one *New Yorker* writer and his wife are welcoming snowbound motorists into their home for shelter and food. It turns into quite a party. When it is all over, the list of guests includes two family groups, a moving van driver and his helper, a mysterious old man who never identifies himself, and six truck drivers, half of whom conveniently were toting nothing but beer.

January 28, 1977. Buffalo, New York. From October through February, storms rolling in off Lake Erie and Lake Ontario routinely dump 1.8 meters (6 feet) of snow on Buffalo. This city of three hundred thousand lies in the heart of a region described by the *Guinness Book of Weather Facts and Feats* as "the snowiest of any populated territory in North America.[138] But on this morning even the city's toughest snowbelt warriors are somewhat taken aback as winds gusting to 135 kilometers (85 miles) per hour sculpt snowdrifts two-and-a-half storeys high on their front lawns. The drifts build up so fast that five thousand vehicles are literally stopped in their tracks. Of the 29 people who die in the storm, many freeze while huddling in their cars or trying to walk to safety through an urban whiteout. Some survivors have to be dug out by rescue workers, the snow having engulfed all but their vehicles' antennas. Entire buildings are buried. Roofs are caving in all over town. President Jimmy Carter proclaims the area a major disaster zone. The US military, aided by massive bulldozers and earthmoving machines, takes two weeks to clean up the mess. Meanwhile city councilors hold a lively debate on whether belching smoke from local factories may have seeded the storm clouds and boosted the snowfall. In *White Death: Blizzard of '77,* Erno Rossi calculates that in just five days "approximately ten thousand square miles of snow blew inland from the surface of Lake Erie."[139] I can't tell you exactly how he came up with this figure but it sounds about right to me. Thanks in part to this megastorm, by the end of the decade enough snow had fallen on Buffalo to bury a ten-storey building.

I'll be the first to admit that urban winters have their downside—that's one reason I don't live in a big city. My idea of winter fun does not include trudging through thigh-deep snowdrifts, white-knuckling through endless traffic snarl-ups, wading through salty puddles, and leaning into frigid blasts that scream through skyscrapered canyons. Major snowstorms can intensify the sting of all of winter's thorns by several orders of magnitude, leading some people to conclude that the end of the world is not far behind. But, as one distraught mayor of Buffalo put it, all storms eventually do blow over, making today's blizzard the "folklore of the future."

While a good snowstorm can choke a city's ability to deliver essential services, it doesn't put a damper on this firefighter's allegiance to coffee breaks.
John Poirier

Hogtown Crumbles under Snow

So, in the urban battle with snow, how far have we come in the 115 years since the Great Blizzard of 1888? In some cities, like Montreal, Quebec, and Charlottetown, Prince Edward Island, applied snow science and the best technology have sown the seeds of faith in "deseasonalization." This means, for instance, that, no matter what falls from the sky, skateboarding down the sidewalk in January should be as easy as it is in June. Other cities lag far behind, usually getting caught with their pants down whenever a big storm hits. Besides embarrassing deficiencies in snow removal technology, one of the biggest obstacles to winning battles against snowstorms in such centers is a chronic and widely held belief that winter does not really exist, not *here* at least. Though economical on paper, this mildly pathological worldview becomes especially troublesome when the Big One hits.

I can think of no better place to illustrate this bizarre urban phenomenon than metropolitan Toronto, Canada's one and only megalopolis. The time: January 1999.

Things began to snowball in Hogtown during the first eleven days of the year when almost a centimeter of snow (0.4 inches) fell each day. Though this doesn't sound like much, many Torontonians were already mildly traumatized. After a one-day respite from the snow, I imagine many people assumed, "Okay, that must be about it for this year's winter." Then, with next to no warning, a feisty "monster storm" from the southern United States came barreling across the Great Lakes and plowed into southern Ontario. According to one CBC news report, wherever it touched down it brought "winter storm warnings, white-outs, snow squalls, brutal wind–chill levels, and record lows." Along parts of Lake Ontario the storm dumped 25 centimeters (10 inches) of snow in 10 hours. One reporter was so mightily impressed by the spectacle of snow-covered streets and stuck cars that, from some undisclosed hat, she pulled out the factoid that Toronto was "very quickly approaching the snowiest January in two hundred years."[140]

Toronto's spagghetified network of highways normally moves about 650,000 people, give or take a few hundred thousand, during your average morning rush hour (read 5:00 a.m. to 10:00 a.m.). Another fifty thousand roll into town on commuter trains. Thousands more depend on Toronto's vast subway system to get to work. By January 14, snow had rendered most of these high-speed arteries dysfunctional at best. In the words of another reporter, "If you weren't driving a snowplow or an emergency vehicle, you weren't driving at all."[141] Even the subway was snowed under. According to one story, "This is how you paralyze Canada's biggest city—cut off its central nervous system—its subway. It's not all underground and what isn't is buried under snow. Its passengers are fed up and furious."[142]

Doing battle in the 'burbs
Northern News Service

Downtown, the battle was not going well at all. Long before the storm was over, Toronto's smallish fleet of snow-suckers was set loose on the streets. These retrofitted hairdryers handily scoop up the snow—that is, any snow less than ankle deep—and melt it in a special onboard blast furnace, then flush the meltwater down the nearest storm sewer. With snow over their axles, storm sewers clogged, and temperatures well below −20°C (−4°F), the snow-suckers vividly proved their true worth, gagging left and right. An impressive fleet of heavy-duty plows, blowers, and graders corralled from Prince Edward Island eventually cleaned up most of the downtown mess.

Somewhere in Toronto's suburban sea, a plow operator doing his valiant best to clear a residential street was presented with a hammer through his windshield delivered by a home owner whose driveway he had just expertly plugged. (I understand the clinical term for this condition is "snow rage.") Meanwhile out at Pearson International Airport, where all the airplanes were allegedly freezing up (funny, they don't do that up north where I live), extra police were brought in from the wings to maintain law and order among fist-shaking crowds stranded by the snow.

A fleet of heavy-duty plows, blowers, and graders from Prince Edward Island helped rescue Torontonians from the storm of '99.

Lorne Schollar

At some point, this kind of snow-induced mayhem drove Mayor Mel Lastman to declare a "snow emergency," something no Toronto mayor had felt compelled to do in sixteen years. During the peak of the storm, Lastman, who incidentally had driven to work that morning, was heard to say, "I'm afraid of tomorrow. I'm petrified of what could happen tomorrow."[143] Ultimately he threw up his hands in defeat and called in the Canadian Armed Forces. Over the next few days, four hundred soldiers reported to the Ontario capital for snow clean-up duty.

What I found interesting about all this was that, though Toronto had indeed received a good dump of snow—more than twice its average January quota before the month was even half over—this was not what you could call an aerial avalanche. Why the big fuss? This question was not lost on several million of my fellow Canadians living anywhere but in Toronto, including CBC-TV's national news anchorman, Peter Mansbridge. For answers, he had the polished gall to confront the Worshipful Mayor himself.

"Mayor Lastman, you know a lot of people in other parts of the country are watching this, and they don't get it. You know, it *is* winter—why is Toronto having such a hard time with this?"

"Well, by tomorrow morning we'll have over 115 centimeters of snow," said Lastman, turning to the camera, "and I don't know of anyone listening who's had over 115 centimeters of snow within less than two weeks." This remark must have got more than a few people chuckling in Maritime Canada, where accumulations of six times this amount are small potatoes. "And they think it's just a little snow," he went on in slightly pleading tones. "It's not. It's a lot of snow. It's more snow than Toronto has ever had in the history of this city, and not only that," he said, turning back to the camera, "the windchill factor—it's so darn cold outside, and every day everything's freezing up."[144]

In the end, 118 centimeters (46 inches) came down on Lastman's constituents that January. Toronto normally spends less on snow removal per person than any other major city in Canada. Over the winter of 1998-99, taxpayers shelled out seventy million dollars—almost double the city's normal budget for snow jobs. That, of course, did not include the army's expenses, like gassing up all the armored personnel carriers that roved Toronto's streets, looking for trouble, with snowplows attached to their noses.

While sympathetic to clogged driveways and frozen pipes, many people tracked the whole show on TV, viewing the winter woes of Hogtown with shameless mirth and merriment. Newfoundland writer and wit Rex Murphy was among them. In a CBC commentary written a full week before the main storm—even *then* the trauma was palpable in the streets—he offered an alternative diagnosis of Torontonians' pathology with snow.

> It's January, it's Canada, it's been snowing; we're in the territory of the desperately self-evident here. . . . However, it snowed in Ontario recently—and some of it fell on Toronto, a fair bit. And judging from the reaction, this is not only unusual, astonishing and a great blathering horror; it's downright unnatural. Even as I speak, the trauma teams are assembling and snow counsellors are on 24-hour call . . .
>
> Unfortunately, I wasn't around when the ancient Egyptians had the ten plagues unloaded on them. But I readily surmise their reaction was very much like that of the great city of Toronto over the last four or five days. Why us, O Lord? Why us? The Egyptians got boils and locusts; Torontonians have Pearson Airport and tow trucks. Very clearly it is not supposed to snow on Toronto; and in the unnatural and inexplicable event that it does snow, sometimes it's allowed to snow for a filmshoot, then quite clearly it's not supposed to stay. This is elementary Toronto physics. In this city, the snow evaporates before it reaches the ground.
>
> Pearson in January isn't really an airport in the conventional sense. It's a great hostel with tarmacs, where snow removal is evidently under the supervision of a consulting team from Fiji. Rule one for clearing the runways: whistle till spring.
>
> Part two of this wonderful equation clearly states that if snow does fall on a busy downtown Toronto street, it is immediately classified as heritage protected. Snow may be allowed to drift, but it must not be pushed; the snowbank is a delicate ecosystem and ploughing is crystal ecocide.
>
> Another wonderful thing about winter in Canada's only megalopolis: every snowfall is a record. . . . If what happened here in the last few days had happened in Winnipeg or Bonavista [Newfoundland], it would not have been called a storm; they would have declared spring break.[145]

Chances are, if you were living out in the sticks or in some two-horse town north of Toronto when that "storm of the century" hit, you could have rolled with its punches in self-sufficient aplomb. The reason for this is simple: the more that complex technology and rigid routines buffer us from Mother Nature, the more punch she can deliver with a few frisky flurries. Such storms forcibly remind city slickers that, despite their cushioned comforts, they always and ever shall be residents of a wilder world, subject to its inscrutable whims and untamable forces. As cities become "more proficient in their services," wrote one urban historian, "they become more vulnerable to [nature's] obstructions"—snowstorms being among the most obstructionist.[146]

In an article starkly entitled "Snow and Man," meteorologist G. A. McKay puts his finger squarely on the root problem of urban snow. He should know. The guy lives in Toronto.

> Not many years ago cars were placed in storage during winter. Today modern society demands that all forms of land and air transportation have virtually unrestricted and safe movement throughout the winter.... In early times a blizzard could isolate a few settlers for days, but today it can paralyze agglomerations of industry, commerce and communication within large, highly urbanized areas....The problems of a snowbound megalopolis far outshadow those of a snowbound town and the massive transport of material and resources over long distances, as required by our expanding society, is increasingly vulnerable to the whims of winter weather ...The hazards created by snow are increasing in number and seriousness as society becomes more populous, urbanized, interdependent, and reliant on transportation. Much technology exists to counter the adverse effects of these hazards, however, all too frequently the problems created by snow are due to a lack of preparation, the snowfall being unseasonably early or late, or of unexpected intensity.[147]

Would-be mayors of Toronto, take note. Expect the unexpected. And, when the next record breaker hits, be prepared to convince a few hundred thousand people to kindly put their cars in storage for a day or two. You might want to start with your own.

Blizzards on the Civilized Fringe

Siberians call it a *purga*. Mongolians call it a *dzud*. The Inuit call it *piqsiq*. A blizzard, by any other name, would sting my cheeks.

How do you know when a blizzard's at your door? When the wind screams and howls outside with an almost human sound. When tiny particles of ice enter your cabin through cracks and keyholes with the velocity of bullets. When your huge malamute dog, chained up outside, tucks into a ball with its furry back to the wailing wind, then promptly vanishes under a thick icing of snow. When you risk your life without a sturdy rope to guide you to and, hopefully, from the outhouse.

Not all storms with howling winds and blowing snow make the grade as a blizzard. Among professional climatologists, an official blizzard meets several chilling conditions. Temperatures at or below −12°C (10°F). Strong steady winds of at least 40 kilometers (25 miles) per hour. Lots of fine driven snow, mostly plucked from the ground. And, thanks to all that snow, visibility of less than a 150 meters (500 feet). There's another condition, one prairie and tundra dwellers are quite familiar with: it's got to last a long time, at least 4 hours.

Taking things to extreme, climatologists declare a *severe* blizzard alert when the winds rise above 72 kilometers (45 miles) per hour and visibility drops to about a centimeter (0.4 inches) in front of your frostbitten nose.

Blizzards rarely begin with a clash and a clamor. They start quietly, slinking toward your door, knocking when least expected, engulfing you almost without notice. Your first clue that a blizzard is headed your way might come the day before when you tap your knuckle against the barometer and the needle takes a sudden steep dive. You step outside. Stretching to the horizon are thick clouds that appear to be fast filling with lead. There is a foreboding heaviness in the air.

A couple hours later you notice a few erratic flecks of falling snow. All seems innocent enough; no tempest here. You stick out your mitt and catch a dusting of crystals and bring them close to your face. The crystals are small hard pellets, rounded and packed by some kind of profound turbulence far above your head. A light evening breeze brings unseasonable warmth from the southwest and false hopes of spring.

While you dream of tundra flowers and returning geese, the tide turns, the tantrum begins. The wind swings suddenly from southwest to northeast, gusting now to 60 klicks (40 miles) per hour. The temperature plunges by ten degrees every hour until it bottoms out at −45°C (−50°F)—and that's not counting windchill. You wake up to a wailing, scraping noise that sounds like everything around you is tearing itself apart. You steal a look out the window and discover a world gone mad with whiteness and wind.

Like snow crystals, no two blizzards are alike. A blizzard may last for days and bring 0.5 meters (20 inches) of fresh snow. Or it may last only a few hours, drastically rearranging the furniture of snow already on the ground. But come what may, it comes fast. Hence the German expression (and the etymological root of the word), *Der sturm kommt blitzartig,* meaning "the storm comes like lightning." Though more plains and prairie dwellers are hit by blizzards than any other North Americans, it is in Canada's lightly peopled eastern Arctic where our fiercest and most frequent blizzards really let loose.

As unpredictable as blizzards may be, long-term weather records from this region show a familiar pattern: at least two or three monster storms a year. Pick any winter. The headlines from the *News North* read pretty much the same: "Eastern Arctic Blasted by Winter Blizzard," "Adventurers Survive Four-Day Blizzard," "Worst Snowstorm in 13 Years," "High Winds and Blowing Snow Hammer Iqaluit." You get the picture.

The blizzard capital of Canada has got to be Rankin Inlet, Nunavut, an arctic hamlet of fifteen hundred souls parked on the climatically unstable shore of Hudson Bay. Anyone who has ever flown there in the winter would agree with me. Most visitors automatically pack at least one fat novel and a deck of cards in anticipation of getting stranded by a storm. Ear-popping winds of more than 100 kilometers (60 miles) per hour, streets blocked by 2-meter (6.5-foot) drifts, phone lines knocked out for hours, power supplies interrupted every few minutes, windows shattered by flying debris, aluminum boats rolling end over end down the frozen shore—the locals have seen it all, and, for the most part, they're ready for it.

"I think people in this area know about high winds and are prepared for it," says Rankin Inlet's mayor John Hickes. "Most people pile insulating snow around their homes. They keep their heating fuel topped up and their cupboards well stocked. Just in case a blizzard hits."[148]

As much as northerners like to stretch their fish stories now and then, they sometimes, oddly enough, tend to understate the savageness of their winter storms. Like the Rankin Inlet man who, digging out after a record-busting blizzard, discovered that his office had been royally flooded when the pipes froze up. All he could say was, "That's the way it goes sometimes when the weather decides to act up." Or how about the school principal who discovered several truckloads of snow on the gymnasium floor? Surveying the indoor snowfield with a local reporter, he said, "I hope this doesn't happen every time we have a little storm."[149]

Believe it or not, somebody actually lives in there. Home swallowed by snow in the arctic hamlet of Cape Dorset, Nunavut.
Tessa Macintosh

A friend of mine cut his northern teeth on the snows of Baker Lake, an arctic hamlet about 250 kilometers (150 miles) northeast of Rankin Inlet which is equally notorious for howling good blizzards. No month of the year is exempt. As I write, in early June 2002, the snows of Baker Lake's most recent "record breaker" are just starting to melt. This one lasted for almost four days, subjecting Baker to gusts exceeding 120 kilometers (70 miles) per hour and temperatures approaching −25°C (−13°F)—remember I'm talking *June* here.

When Damian Panayi was seven years old, his family moved from the virtually snowless town of Parksville, in the subtropics of Vancouver Island, to blizzard-battered Baker Lake which, as many of its thousand or so residents will tell you, is situated at the true geographic heart of Canada. While sitting around a campfire one winter day at my River Lake cabin, eating corn soup and bannock, I asked the now adult Damian about the celebrated blizzards of Baker.

"When it snowed in Parksville—just a skiff like this," said Panayi, kicking a fresh dusting of snow into the fire, "my parents always kept me inside. It maybe got down to −5°C [23°F] once or twice in my memory. By the next winter we'd moved to Baker Lake, and I was playing street hockey at −35°C [−31°F]. The snow in Baker was totally amazing. It regularly drifted up and over our house, so high we could run from the street to the roof and jump off the other side. We usually had to duck under the telephone wires on the way up. My dad had to shovel snow away from our front window just to get some daylight into the house."

Twenty years later, Panayi made his first winter visit to the subarctic, where I live, and my family took him on a picnic to the local waterfalls at −30°C (−22°F). This temperature was nothing, of course, to an arctic resident like him, but it turned out to be his first immersion experience in taiga snow.

"I couldn't believe how fluffy it was. It didn't make sense—to be sinking into snow. Back home in the Arctic, the snow gets so hard you can jump on it and not leave a bootprint. Shoveling the stuff is no laughing matter."

Up here, people like to debate about what it takes to become a bona fide Northerner—please note the capital N. Besides swallowing ice worms or seeing the so-and-so river break up, the number of winters spent north of sixty always seems to fall somewhere in the equation. Civil servant and writer Duncan Pryde spent ten years above this latitude and gathered at least enough northern credentials to survive a mighty blizzard that struck beautiful downtown Baker Lake in 1970.

One of the first things the local Inuit taught him was to carry a snow knife at all times, even around the settlement, just in case a blizzard struck. The idea was that even if he couldn't build an igloo, he could at least carve a windbreak in the rock-hard snow. The other thing they taught him was that an arctic blizzard is dangerous only if you panic and, wrote Pryde, "when you panic, you're halfway dead." In his book *Nunaga: Ten Years of Eskimo Life,* Pryde tells how these two tips, and a pinch of luck, probably saved his life— just a snowball's throw from warmth and safety.

One night at Baker Lake a friend working at the meteorological station and I accompanied two young school-teachers to the movies at the [Department of Transport (DOT)] station. The weather had been calm in the early evening, but by the time the show had ended a furious storm had completely cut off visibility. A power cable ran all the way from the DOT building to the schoolhouse where the girls lived; we thought we could easily follow it. But in the flurry of the wind we not only lost the cable, we got lost ourselves—right in the middle of the settlement. I wasn't particularly worried; I was dressed from head to toe in my [caribou] skins, but the others were wearing ordinary white man's clothing. . . . They were frightened. The wind drove hard snow granules into our faces like ice bullets, and it felt as if it was cutting the flesh of our cheekbones away. One of the teachers panicked and had to be physically restrained from dashing blindly into the darkness. Yet we might as well have been blind. All we could hear was the moan of the wind and our own laboured breathing. The dogs, dug in against the blow, were unusually silent. We couldn't have been more than a hundred yards from the Hudson's Bay post, but no buildings were distinguishable. We could have been a hundred miles out on the tundra.

The white man's way is to keep walking in a blizzard to try to keep warm. But the Eskimo will not try to fight the elements and tire himself out. The Eskimo way is to burrow into a snowdrift and let the snow itself protect and insulate him from the bitter wind. I proposed this, and the others agreed. I held one girl in my arms and the other fellow held the other girl. We got them quieted down, and it seemed not a bad idea after a while. . . .

But we were like most white men—no patience—and by now, incapable of the state of mind that would allow us to make use of the very elements we fought, as an Eskimo would. We elected to venture out and luckily stumbled into an Eskimo shack. The inhabitants told us which way to go to reach the school. Even then, though the school was less than a hundred yards away, we almost missed it.[150]

In spite of the best precautions and the most level of heads, blizzards and tragedies still go hand in hand. According to David Phillips, a federal climatologist, "Winter storms and excessive cold claim over 100 lives in Canada every year, more than the combined toll from hurricanes, tornadoes, floods, extreme heat and lightning. Blizzards are easily the most feared and perilous of winter storms."[151]

As they have for centuries, even the Arctic's seasoned hunters occasionally become stranded on the land by a blizzard so fierce that navigating or building a snow shelter becomes impossible. Ambushed by the white wind of a blizzard, they die.

Gale-force winds fill the streets with snow on Holman Island, Northwest Territories.
Tessa Macintosh

But as wild as our winter storms get, no place experiences blizzards like the ones that assail Antarctica, one of the windiest places on Earth. In 1934 Richard Byrd, the famous aviator who was first to fly over the North and South poles, spent a long winter alone, burrowed in a bunker beneath the antarctic snow. Here's what he wrote after finally digging himself out.

> It is a queer experience to watch a blizzard rise. First there is the wind, rising out of nowhere. Then the [land] unwrenches itself from quietude; and the surface, which just before had seemed as hard and polished as metal, begins to run like a making sea. Loose crystals are moving as solidly as an incoming tide, which creams over the ankles, then surges to the waist, and finally is at the throat. I have walked in [blowing snow] so thick as not to

be able to see a foot ahead of me; yet when I glanced up, I could see the stars shining through the thin layer overhead. . . . The air came at me in snowy rushes; I breasted it as I might a heavy surf. No night had ever seemed so dark. The beam from my flashlight was choked in its throat; I could not see the hand before my face.[152]

Flashlight or no, Byrd discovered that there was little point in opening his eyes. "To see was impossible. Millions of tiny pellets exploded in my eyes, stinging like BB shot. It was even hard to breathe, because snow instantly clogged the mouth and nostrils."

If you live in blizzard country, you have to expect the unexpected and roll with the punches. Take the whopper blizzard that hit Iqaluit a couple summers back, right in the middle of the town's Canada Day celebrations.

"It was July 1," said a woman who lived through it. "We were having a big party—barbecue, balloons, the whole bit. Then a blizzard hit and everybody had to move inside. We made the best of it."[153]

Making the best of a blizzard is something northerners (however defined) are particularly good at. We generally live with the roughest edges of nature more or less in our faces. If now and then we forget this, a gutsy blizzard quickly brings us to our senses, forcibly demonstrating our vulnerability to the elements. When our civilized comforts lull us into relying more on technology than our land-based wits, when we occasionally fall prey to the delusion that we are nature's masters, blizzards offer a rude slap in the face. On the other hand, if we are physically and psychologically well prepared, blizzards offer a unique communal experience in northern living.

Viewed from within a warm, well-stocked house, blizzards can be more cause for delight than alarm. As the wind pounds at our door and the radio announces the closing of schools and offices across town, we get a welcome reprieve from our daily chores and responsibilities. A time-out. We break out that novel or deal out those cards. We catch up on our sleep like hibernating bears. Or we merrily eat humble pie with family or friends, while the blizzard rages without.

The Technical Fix

When it comes to making snowy cities more livable, there seems to be two polar camps. The right-wing chionophobes want to control snow, stomp it down, if not get rid of it completely. More on them in a moment. Way over in left field is Washington's Frederick Gutheim, a radical urban planner-cum-philosopher who unveiled his somewhat flaky ideas about urban snow in a 1979 *Architectural Digest*. He suggested, for one thing, zoning northern cities to allow snow to remain untouched in certain neighborhoods "for aesthetic effect and recreation." He's so crazy about snow that he even proposed *making* snow and blowing it all over the streets, parks, city hall, and who knows where else "to brighten their appearance and encourage nonmotorized transportation."[154] If he ever tried that in Toronto, I would strongly encourage the guy not to run for mayor. Still, he may be on to something.

For those chionophiles who would join Gutheim's camp, I'm happy to tell you that without resorting to political pressure, pleading, or threats, you can go out tomorrow—if it's cold enough—and crank up the volume of snow that lands on your town. I'm not talking cloud seeding here. That's a tad controversial. The place to start is your own backyard. Thanks to the inventive mind of Massachusetts resident Charles Santry, you too can thwart a brown Christmas with a personal snowmaking machine called—what else?—a Backyard Blizzard.

Santry and his partner Doug Harvey fell on this idea quite naturally. Their livelihood is building giant snowmaking machines for ski resorts. "The challenge was making the Backyard Blizzard into something you could use at home," says Harvey.[155]

About half the size of riding lawnmower and emblazoned with a snowboarding polar bear, this intriguing piece of technology is quite simple to operate. All you need to cook up a storm is a twenty-amp outlet and an old garden hose. In an hour you can blanket about 37 square meters (400 square feet) with 2.5 centimeters (1 inch) of snow—the colder and drier the weather, the better. Lay down two thousand US bucks, plug it in, and let 'er rip, all over your empty swimming pool, your neighbor's lawn, or whatever else strikes your fancy. Get creative. In Saudi Arabia, for instance, they're using these things at weddings.

In an online advertisement, Santry includes a field testimonial from a loyal Backyard Blizzard user, one Morgan Wesson of Rochester, New York (as if they didn't get enough snow there already). "It just cooks," says Wesson. "At first you're real nervous." The ad stops short of telling us exactly what Wesson is afraid of. His neighbors perhaps? "You stand there and start looking at the tower and down at the pressure gauge and the kids are dancing around! It just purrs along and keeps pumping this stuff out. . . . Until it blows across to the neighbor's yard."Yes, it must be his neighbors.[156]

On the other hand, perhaps you fall into the camp where the subjugation and removal of snow is the main thing. At the outer fringes of this camp are those who would solve all the headaches and hazards of urban snow by not letting it pile up in the first place. These are the dome-heads, Buckminster Fuller being the domiest. I once heard this renegade engineer-inventor and Harvard dropout speak when I was a freshman at university. Most of what he said was undiluted Greek to me, but I do remember something about domed cities and snow.

I didn't give Buckminster Fuller another thought until many years later when I came north and was introduced to my first real boss, Arthur Boutilier, a Bucky groupie if there ever was one. From the ceiling of Boutilier's office hung a menagerie of colored Styrofoam spheres—"Bucky balls," he called them—arranged in tetrahedral splendor. These apparently were the mystic cells that gave strength and shape to Fuller's skintight geodesic domes. From day one, my office in-basket regularly groaned under weighty towers of Bucky literature, personally delivered by Boutilier, which I assumed had some vague connection with my new job as an environmental planner. I dutifully dove right in.

I learned that Fuller once proposed, in all seriousness, stretching a dome over a large section of Manhattan, arguing that he could take the sting out of its notorious winters while saving the city millions of dollars in heating and snow removal costs. Perhaps he got the idea while browsing through back issues of the *New York Times* where, after a grand storm in December 1879, one reporter wrote:"The evil snow is upon us. . . . A century hence, cities will be put under glass and New York will be enclosed in a huge crystal palace."[157]

Though Fuller's plan to zip up Manhattan soared like a lead balloon, it caught the eye of free-spirited planners in St. Louis, who hired him to spruce up the city's slummy east side. Here Fuller pulled out all stops, drawing up detailed plans for a domed community of 125,000 nestled in a crater-like valley. Dubbed "Old Man River's City," it was overarched by a transparent "geodesic sky parasol umbrella."[158] The dome measured over a 1.5 kilometers (1 mile) across and was designed to protect its proud and happy citizens from wind, rain, and, of course, snow. After much media hype and political hoopla, this plan too was consigned to the dustbin of history.

Digging deeper through my teetering in-basket, I also learned that in the late 1950s, an architect infected by Fuller's outrageously practical ideas developed a detailed plan for doming over much of Iqaluit, then called Frobisher Bay, now capital of Nunavut on Baffin Island. His idea generated at least enough momentum to adorn the pages of a 1959 issue of *Popular Mechanics*. The article showed a futuristic ring of high-rises encircling a central domed common area and claimed that the proposal was "under consideration by the Canadian government."[159]

Apparently the Canadian government is taking its sweet time on this one, for Iqaluit remains as yet undomed. But as an ardent Bucky-phile, Boutilier remains ever-faithful to the idea. In the arctic spring of 1980, working as a Northwest Territories community planner, he had a powerful vision while gazing down on Kimmirut, a tiny Inuit settlement rimmed by high cliffs that form a perfect bowl opening to the sea. "We were standing on a high point," he recalls, "and I thought 'Wow!' You could just visualize how to enclose the community."[160]

His idea was to stretch over this bowl a tentlike skin, based on Fuller's design principles, that would yield maximum strength and protection with minimal materials. "You could take away the harsh [winter] climate and have the same mobility you have in summer." Needless to say, Boutilier's government masters labeled his rough pencil sketches of a parasolled polar hamlet as "pie in the sky" (they were not, in fact, far off the mark). "We were ahead of our times," muses Boutilier. "It wasn't to be. But," he hastens to add, "the reasons for building [domed communities] will never disappear."

It turns out that if you want to see anything on the ground that faintly resembles Fuller's wildest dreams, you're in for a long plane ride to the South Pole. There you will find a complex of research buildings nestled under a snow-shedding Bucky dome. If you were alive and paying American taxes in 1975, you may be happy to know that you helped pay for this radical yet eminently efficient structure that now cocoons the Amundsen-Scott Research Station. Whipped together by a platoon of Navy Seals in just a few weeks, the 45-meter (150-foot) -diameter dome is draped over three two-storey buildings housing about sixty people. Though not exactly a metropolis, imagine the diameter of Bucky's eyeballs had he done a South Pole pilgrimage and witnessed firsthand one of his beloved domes standing unflinchingly up to the wildest winter weather on earth. Fuller might, however, have paused a moment when he stepped inside the compound to discover snow below his mukluks and around the buildings. The structure itself, though relatively balmy inside, is purposely not heated. It wouldn't take long for a genius like him to see the sense in this. You'd have to melt about 1.5 kilometers (1 mile) of glacial ice before settling your buildings on anything resembling solid ground.

After more than 25 years worth of blinding blizzards and −60°C (−76°F) temperatures, Bucky's brainchild sits unscathed in a wasteland of perpetual winter. The only design hitch is that all that blowing snow has sunk the entrance to the dome station right out of sight. The only way in is through an unplanned tunnel. Okay, so you try coming up with a better design for a blizzard-beaten place like this. I wager you won't find one. If it works down there, why not in your home town?

As recently as 1990, city planners in snowy Montreal were blowing the dust off Fuller's ideas. Mayor Jean Doré himself, while addressing a cosmopolitan crowd of Winter City conference delegates in Norway, posed the question, "In winter cities, is the final target to use scientific technology to shut out winter by enclosing the city under a huge dome?" Such a rash proposal would not have surprised his listeners, knowing that his city takes great pride in its "defeat and dispose" approach to snow removal. Using military-like tactics and technology, Montreal's strategy, on paper at least, aims to keep all roads and sidewalks cleared all winter. But no, says Doré, sour-graping the dome idea with a shot of Darwinian theory. "Isn't it winter snow and coldness which gave us the driving force to develop into what we are now [as a species]? . . . We should not escape from winter; we must not deny its presence. In this sense, it is time to reconsider the application of scientific technologies."[161]

Darwin aside, maybe Montrealers aren't hankering for a dome, since they already have more than 900,000 square meters (10 million square feet) of subterranean shopping malls. Their snowproof subway system is known as "the largest underground art gallery in the world." Minneapolis recently toyed with the idea of putting a lid over two city blocks, while Thunder Bay, Ontario, honestly considered roofing over its main street. Many other urban centers in snow country have creatively covered up for winter without resorting to full-blown domes. But that shouldn't stop you from going for it.

Instead of aiming a personal snowmaking machine at your backyard, I suppose the technological antithesis might be to cocoon the whole thing— swings, sandbox, go-cart track, whatever—under geodesic Plexiglas. Maybe slip your house and garden under it while you're at it. But barring domes of any size, the reality is that you're probably, at some point or other, going to have to shovel the stuff off your driveway. At this prospect you may rather, as one *Financial Post* writer put it, "pass a kidney stone in a warm and comfortable bed than get up and lift shovels of snow weighing five to seven pounds each."[162] And what semi-able male over forty does not think about a heart attack or a thrown back with every toss of snow?

Do you check your pulse with every shovelful? Keep the whole body dancing while you dig and such thoughts will not trouble you.

Northern News Service

But fear not. The ergonomic shovel has arrived. I own one, in fact. Fresh from the shelf it appeared to have been run over by a large moving truck. Looks aside, I can now attest that there's something in its kinky handle and swoopy blade that magically lightens the load. Somebody was thinking. Maybe Bucky had a hand in it. If you view snow shoveling as Tom Sawyer treated whitewashing his fence—a chore to be graciously palmed off on others—you might instead choose the Oskar model. This ingenious device has a telescoping handle with over 0.5 meters (1.5 feet) of play, enabling you to recruit all members of your family, regardless of shape or size, in the driveway assault. For those with ample reserves of brute strength and superhuman patience, there's the Poly Sleigh with a gaping maw that gulps about 110 cubic decimeters (4 cubic feet) of snow in one pass. Designed not to lift snow but to push it, all that is needed to operate this behemoth of a shovel is a human bulldozer. Finally there's the good old Nordic, which, for less than ten bucks, is a standard plastic workhorse designed for your average Joe shoveler who has a trustworthy heart, a strong back, and no money.

If you are indeed absorbed in passing a kidney stone or otherwise obliged to turn your sore back on snow shoveling, there is always the snow-blower route. Here, too, there are wide options for the discriminating 21st century snow warrior. According to *Consumer Reports,* when it comes to snowblowers, "horsepower matters."[163] Also, did you know that there are such things as single-stage and two-stage blowers? You didn't? Get with it. (I didn't either until I looked into this.) Apparently the singles work fine on snow up to 20 centimeters (8 inches) deep, then they sputter and choke. The two-stagers will devour up to 45 centimeters (18 inches), no problem. Beyond that, I guess you're stuck.

If you live in Victoria or Seattle and want to impress your neighbors the next time (or year?) it snows, but don't want to fork out the dough, you might consider a four-horsepower single-stage midget. This model will do a grand clearing job of your back patio—or you could just wait a few hours until it all melts. For the middle-roaders there's a two-stage, seven-horse-power Ariens, described by its builders as an "upscale residential model." They designed this machine "for somebody who's kind of in a hurry in the morning and wants something reliable."[164] With a six-on-the-floor stick-shift, two backwards speeds, and a price tag of fifteen hundred dollars, this baby promises hours of modestly priced entertainment even when there's no snow to blow around. If caught in the act, you could tell your neighbors you're just doing training exercises for the next upscale blizzard.

If your driveway has the dimensions of an airport runway, you might consider dropping close to four thousand clams to buy a fifteen-horsepower model that can carve a cool clean swath more than 90 centimeters (3 feet) wide. These macho machines have the guts and teeth to tackle just about any snow in sight with as much effort as it takes to carve warm butter with a chain saw. Lose yourself in the unbridled power at your mitten-tips. Revel in the great geyser of snow that follows you everywhere. Exult in the altruistic joy of serving your unfortunate neighbors still buried in snow by clearing their sidewalks or maybe even their driveways. While you're at it, you might as well tidy up the edges of your street left ragged by a speeding snowplow. Most of the bigger snowblowers are furnished with a motorcycle-sized gas tank, hand warmers, and dual high-beam headlights. Thus equipped you could, if you really got into it, keep this up all night—or at least until your neighbors call the cops. These big ones sound like a small army tank that's lost its muffler.

If your cardiologist has put you on salt-free tofu and your chiropractor is getting rich quick, maybe it's time you called the whole snow clearing thing off and hire a team of commercial experts to dispatch any white stuff that's in your face. There's an outfit in Toronto, for instance, that guarantees a snow-free driveway all winter long for around three hundred bucks. For another hundred or so, they'll clear all your sidewalks, pathways, steps, even shovel your roof if it's in danger of falling in on you. Such fees soar quickly upward with lot size. If you happen to live on an estate lot in Toronto's Rosedale or Bridal Path district, you will pay through the nose, signing checks approaching three grand for the privilege of watching Bobcats and dump-trucks cart away your snow. You have to wonder what these guys would charge if they operated out of Baker Lake.

The ultimate in personal snow removal technology is, of course, the push-button driveway heater. Just imagine how you'd feel watching your neighbors quarrying through piles of heavy wet snow, while you sip herbal tea as the snow in your driveway passively melts and trickles away. (Personally I would feel like a schmuck.) For the sake of your health, safety, prestige, or some primal lust for conquering nature, you might choose this route, deciding that its benefits exceed the nominal installation cost of about twelve thousand bucks. Keep in mind, that's not including the cost of electricity to power an underground toaster spread over a typical driveway's 80 square meters (860 square feet) of asphalt. If you ask me, you'd really have to hate shoveling to go this far.

Just how much you are willing to pay to get rid of your snow may depend on the number of expensive toys already sitting in your garage, or, should things go wrong for you out there on the driveway, the adequacy of your medical insurance coverage.

Me, I shovel my driveway maybe half a dozen times a winter, sometimes for looks, mostly for exercise. I do a strident swinging two-step with each scoop to engage as much of my body as possible and keep the blood safely flowing to all quarters. I don't own a garage—nobody does in my neighborhood—so just parking the car in my driveway keeps most of it clear all winter. The only downside to this strategy is that as snow builds up around the car, it creates a bowl-like snow shadow into which the car irrevocably sinks over the winter. This phenomenon was driven home to me forcibly after a particularly heavy snowfall. While removing groceries from the rear of my car, my skull struck the open hatch which was in effect several inches lower than before it snowed. The pickles and pasta took flight, jets of blood painted the snow, and my mouth issued severe reprimands to the offending tailgate. But everything turned out all right. My medical insurance covered all six stitches.

The Mental Fix

Whether you live in Baker Lake, Nunavut, or Buffalo, New York, the ultimate weapon in the fight against urban snow is the same: a healthy and humble attitude. The flip side—the attitude that guarantees trouble when the storm hits—is the notion that we can carry on as if all that snow did not exist. Deny its hazards, and "business as usual" spells a recipe for disaster. Slow down, surrender to the snow and discover that there is life beyond the treadmill. You might even live longer. Nowhere is this truism put on grander display than in sprawling high-strung cities where snow is an uncommon and generally unwelcome guest.

A former neighbor of mine, Chris Straw, packed up one day and moved to Vancouver. He later told me that what he misses most about northern winters is that "cornflakes crunch of dry snow under your boots when you walk to work." I happened to give him a call on the very morning a snowstorm engulfed his adopted city of half a million, in the merry month of May, months after the spring daffodils had first shown their pretty little heads.

"Believe it or not, it snowed today in Vancouver!" Straw told me with much amusement. Weather is indeed unpredictable in Vancouver, hence the saying: "People from Saskatchewan move to Vancouver because of the climate; they move back to Saskatchewan because of the weather." But snow in Vancouver in *May?* Impossible.

"I live in North Vancouver, below the foot of Mount Seymour," said Straw. "We're maybe twenty feet above sea level, and I woke up to snow on my front lawn. They make a lot of movies around here, and I honestly thought somebody was pulling a fast one on us for the sake of some production that called for snow. People were freaking out."

I asked him to be a little more specific.

"Everyone piled into their SUVs like everything was normal, whipped them into four-wheel drive, then started slamming into each other. It really was pathetic to watch."

How people drive when the snow starts flying provides a good barometer of trouble ahead—much of which is avoidable if only they'd slow down or simply bundle up and walk.
John Poirier

How people drive their vehicles when the snow starts flying seems to be a good barometer of the trouble they're in for, much of it self-inflicted. Ted Warren is a retired building contractor who lives high on the

Scarborough Bluffs overlooking Lake Ontario. From a palatial wooden deck that he built himself, he can watch pileated woodpeckers drumming on his maple trees, white-tailed deer darting through a cedar swamp, and the odd snowstorm rolling in off the lake—headed straight for the soft underbelly that feeds downtown Toronto: a gargantuan freeway system rivaling that of Los Angeles. While on a wildlife tour of the Northwest Territories, Warren described to me the foibles of Toronto drivers who don't believe in snow.

"If a centimeter of snow falls, they'll declare a city-wide emergency," he told me. "An inch of snow is a downright disaster. Everyone is used to cruising at 120 kilometers an hour—it's a God-given right. It wouldn't even occur to most of them to slow down, even when their windshields are getting plastered in brown slush. It's not that they won't slow down. I honestly think they don't know *how.*"

Even in Toronto's horse-and-buggy days, when great dumps of snow were more familiar, its citizens needed the occasional pep talk from the pulpit to get them to look beyond the trials of urban snow. After a weeklong tempest in January 1896, Reverend Thomas Sims fleeced the winter blues from his Bond Street congregation with his now classic sermon, "The Treasures of the Snow." Commuters, take note. Reverend Sims had the option of leaving his buggy in the barn and walking to work that morning.

> As we came to church last Sunday morning, all the slush and grime of the city was buried beneath a thick canopy of purest white. The air was soft and all the land was still. Earth and air and sky all spoke of purity and peace. Whether it was the sweet, fresh beauty of the scene, or some half-conscious reminiscence of early youth, I cannot tell, but, to me, that walk to church was a benediction. . . . Yesterday and the day before we had another. I have gazed upon palaces of queen and noble, I have stood in awed admiration before the sculptured facade, and in the pillared aisles of gorgeous cathedrals, but I never saw in any work of man such stupendous magnificence as Jarvis Street presented on Friday and Saturday mornings. Beneath was the snow carpet, into whose deep pile one's feet sank in luxurious silence. Out of it rose the pillared splendour of chestnut, elm, and maple like columns of alabaster stained by time. Starting from their capitals was an exquisite elaborateness of tracery, which architecture at

its best begins feebly to imitate. The double rows of trees, inter-
lacing over the sidewalks and overarching the street, formed the
gorgeous transept and glittering aisles of a cathedral worthy to
be called celestial.[165]

In *Snow in the Cities,* Blake McKelvey describes the dawning days of
urban North America, telling story after story of "light-hearted citizen
responses to heavy snowstorms." Instead of cabin fever or highway hysteria,
a good snowfall was more apt to kindle "merriment and mutual visiting"
among housebound families and neighbors. Because of snow's peculiar
power "to cultivate the social feelings," the deeper it lay, the deeper early city
folk would reach into their pockets to help the urban poor. Or the quicker
they'd reach for their shovels to dig out an elderly neighbor. Freed from
everyday work routines, they'd harness up the cutter sleigh or dust off the
snowshoes to explore their transformed towns. Strengthened family bonds,
closer-knit communities, more charitable hearts, improved fitness. These,
too, blew in with the storms of old. McKelvey concludes that today's urban
chionophobes have much to learn from history. "If the early cities set an
example worth noting, it was in their capacity to appreciate and find delight
in a snow-covered environment."[166]

Face it. History decrees that no amount of fancy technology or urban
fortification will ever dampen a snowstorm's power to give our cherished
daily routines a thorough drubbing. And here's the thing: in most cases, it's
our attachment to those routines that traps us, not the snow itself. Don't get
me wrong. I like an orderly day as much as the next fella. But I side with
McKelvey when he tells us we can find a silver lining in every winter storm
cloud—if we slow down enough to look for it.

> In spite of all urbanized efforts to predict, combat, exploit, and
> otherwise cope with heavy winter snows, the recurrent blizzards
> that sweep over many northern cities *still* possess awe-inspiring
> force and often leave a glistening landscape that can briefly
> inspire snow-bound residents with a renewed sense of commu-
> nity with their neighbours and a revived affinity with nature.[167]

When Washington, DC, disappeared under the 1979 "President's Day
Snowstorm," an editor from the *Washington Post* eloquently summed up the
futility of complaining. "The message [of the storm] was fairly clear . . . a

Canada's capital disappears under snow, reminding us of a higher seat of power.
John Poirier

knock in the head to the human pretensions of controlling one's own des-
tiny, that whole I-am-the-master-of-my-fate business."[168]

Eight years later, when a single January storm dumped 35 centimeters
(14 inches) on the capital, a *Post* reporter who garnered public feedback on
the event wrote that, "in some ways, being stuck in the Great Blizzard is like
forced psychotherapy on a massive scale."[169] Exactly what the fruits of such
therapy were he didn't say. In my books, the best mental fix is the realization

that, when you wake up to snowdrifts tickling your eaves-trough or erasing your street, sometimes all you can control is your attitude. When snow throws a damper on your day, why not enjoy it?

The next time a whopper storm hits your neighborhood, park your buggy or SUV for a day or two, embrace your professional dispensability, and kick back in the snow. Revel in that cozy comfort that comes from being tucked in a warm home after a storm. "Snow makes me feel as snug as a vole," wrote journalist Cullen Murphy in his 1995 *Atlantic Monthly* article, "In Praise of Snow." During one winter alone, he became thoroughly snowed in ten times after a series of major snowstorms all but shut down his region of New England. Freed by a healthy and humble acceptance of the storms, he "contentedly wrote a check to the snowplow man each time."[170]

Like me, Murphy probably does most of his writing at home. I've worked from my house for a dozen long winters and have never been late for work. To sweep out the night's cobwebs, I leave the house every morning at 8:00 a.m. sharp—I told you, I like routines—walk a secret course of crooked streets, then return refreshed to my place of work to tap away at some momentous writing project, or, for instance, to record a snippet or two in my snow journal about a particularly fine walk one dark January morning.

An ancient road warrior subdued and silenced by the snow.
John Poirier

My cheeks and temples sting from the cold during an early morning walk down Ragged Ass Road. A howling snowstorm carried on most of the night until the clouds parted, the temperature dropped, and the wind died. Besides [my dog] Haley, I am the first to track through these pristine crystals now sparkling orange under the streetlights. I thought it would never come again: −40°C. At last, that frigid, friendly time of winter when the town lies steamy and still under piercing stars and planets. This road becomes a wide walking path where all the cars sit idle, subdued and silenced by the snow.

Snow's Final Punch

The best time to test your mettle as far as snow is concerned is when you least expect it—long after winter is supposedly dead and buried. Somewhere on this good earth, perhaps only on my calendar, winter is officially over on March 21. Calgary writer and adventure-gear seller Mike Cormier believes this date requires some serious revisiting. In the spring 2001 issue of *The Monthly Mountain,* he insists that "anyone who thinks that Spring begins on March 21st needs to have his head examined. Or, more specifically, any Canadian who thinks we've seen the last of snow by that laughably premature date needs to have a nice lie down on the doctor's leather couch and give those old ink blots a little look-see. March? Spring in March? That is not spring." As a subtle "for instance," Cormier draws our attention to the fact that "Calgary received roughly 15,000 times more snowfall in the month of April than it did in the month of January." Trusting in government-supplied statistics, he notes that January's snowfall was so minuscule it was impossible to measure, "although drifts of approximately 5 microns were measured in some outlying areas." The Paul Bunyan-size dumps that followed in April gave Calgary a very unspringlike countenance.[171]

To Cormier the solution to all this is simple: assign different spring arrival times to different regions. "If it is to be the psychological boost we all want it to be, then surely we've got to pick a better date. The first day of spring in Victoria would come some time in late January. In Toronto, it could be . . . aah, who cares about Toronto? And in Calgary, it comes in mid-April. April 15. How does that sound to everybody? I'll tell you what it sounds like, it sounds like the first day of spring."

Great idea, Mike, but did you happen to look out your studio window last May? If not, did you maybe check the papers? I have in front of me the front page of an *Edmonton Journal* showing a young mother towing her toddler through knee-deep snow down the center of a street clogged with abandoned SUVs. The date is May 7, 2002. The headline: "Late Storm Delivers 40-cm Drubbing, with More in Store." The urban fallout reads like a typical December storm: crippled highways, stranded commuters, grounded flights, closed schools, dead snow shovelers. Downtown pedestrians were dodging chunks of metal cladding, stripped from office buildings by leaden snow. And out on the local turnpike, more than one hundred fender benders culminated when two Calgary firefighters smacked into a semitrailer truck while responding to an accident and had to be cut from their own mangled vehicle. Said the police officer who carved them out, "When you have to start rescuing your own, it's not a good day."[172]

Meanwhile Alberta's prairie farmers, just rising from hibernation, were rubbing their hands with glee. They welcomed the extra precipitation and cooler temperatures in the hopes of a gradual moistening of their parched farmlands. "From the perspective of the prairie people," said one agricultural spokesman, "it's been a very good spring."[173] As in the bleak midwinter, the worth of spring snowstorms seems to lie very much in the beholder's eye and pace of life.

Given a vote, I'm not sure on what date I would peg our first day of spring. Up here, spring is a state of mind elicited by the laughing of red-throated loons, the exit of stars from the night sky, and the popping of pussy willows in my front yard. If pressed, I would say these threads in the season's fabric converge around May 15. As for snowstorms, I've come to expect them any time that month. They are always welcome at my door.

The local media greeted a recent May storm with headlines like "Winter's Last Stand," "Dirty, Dangerous Spring," and "When Will It End?" Not one of the articles described how much fun it was. Instead, I read about "record-busting windchills," cabin fever victims, and "sky-rocketing sales" in parkas—you know, the usual stuff. I can't blame the reporters for such knee-jerk coverage; most of them are recent transplants from the south, probably Toronto. Besides, "brutal blizzards" sell newspapers. While they clanged on about catastrophes, here's what I was writing in my snow journal about the same storm.

May 2. Hurray! A spring blizzard is shaking the house. I dial up the official Environment Canada weather robot. Its ET tones always remind me how much I miss the human meteorologists who used to pick up the phone. When prompted, I hit 3 for "weather warnings." A metallic voice feeds me data in chopped-up phrases. "Severe weather warning continues for Yellowknife, Great Slave Lake, and the entire Mackenzie Valley region."

I hit 1 for the forecast. "More snow and blowing snow. Expected accumulation overnight: 15 centimeters. Winds north 40 kilometers per hour gusting to sixty, giving a windchill of −26°C." Now, that's something for May. "Winds diminishing by morning, then shifting to south and rising to 40 kilometers per hour with possible gusting to sixty."

I clap my hands together at the news. A two-for-one blizzard coming at us from opposite ends of the compass! Just in case we didn't get the message: spring is late; winter still rules.

I bound upstairs to look at my neighbor's wind sock, mounted on top of a crooked flag pole in the middle of his yard. The fish-shaped sock is barely moving—locked sideways and stiff as it gulps gale-force winds roaring off the frozen lake. For a moment, the house across the street disappears in a frazzled cloud of wind-whipped snow. Later I watch [my ten-year-old daughter] Nimisha do the same as she runs out onto Primrose Lane to greet the storm with raised arms.

I run downstairs and look out another window to check the thermometer. At first it appears that the thermometer has bottomed out at −45°C. I laugh right out loud when I realize that the blizzard has plucked its red alcohol tube clean away.

You wouldn't think that any winter going out like a lamb would capture much media attention, certainly not on page one. No catastrophes in that. But at the end of an exceptionally mild winter in 1987, the St. Paul *Pioneer Press and Dispatch* carried a front-page story claiming that their weird weather had triggered a serious identity crisis in many Minnesotans. The season's tepid temperatures and dearth of snow had created a loss in the sense of "who they are." One popular weatherman went so far as declaring that, "This [mild winter] is causing severe emotional damage. It's sort of like the Vikings [football team] moving indoors to play." One resident quoted in the article discovered a new neurosis during the winter-that-wasn't. "We get restless, we get inverse cabin fever. We can go out and we don't know what to do."[174]

Psychology professor Jeffrey Nash of Macalester College had been studying the behavior patterns of people in this region over the previous six winters. His results pointed to the positive spin-offs of urban winters, most vividly appreciated when conspicuously absent. "The accepted tenets of urban life—the distrust of strangers, the lack of community, the *strictly* defined use of public space—all seem to melt when winter approaches," wrote Nash. "Winter lessens all that and makes life in the city more like life in a small town. A long winter becomes a common experience that brings people together. This year it just wasn't there."[175]

Severe climate and winter tend to foster the blossoming of urban art, sometimes of the most unexpected kind. Christmas globes and twinkling lights bedeck this snowbound chopper.
John Poirier

In a 1989 interview with the *New York Times,* pop meteorologist William C. Rogers went even further by declaring that long, cold, snowy winters bring out all that is good and glorious in urban life. Much as pregnancy should not be viewed as a pathological disorder, Rogers chided "winter cities" such as New York, Denver, and Chicago for viewing winter as nothing more than an irksome interlude, "something to be gotten over with," rather than as a defining element in their existence that is worthy of special regard and celebration. Though the *Times* lampooned Rogers's comments under the headline "Ah, for a Bone-Chilling Icy Winter," it granted him generous column space to air his iconoclastic views on winter, the wildest of which—though I tend to agree with the guy—was that "severe climate and winter tend to foster culture. The great symphonies, the great art museums are, for the most part, to be found in the winter cities." Rogers's upshot was that "culture thrives better in the cold."[176] Case in point: where are the biggest Disney strongholds, the antithesis of what Rogers is talking about? All in the steamy south, where snow is seldom seen if even believed in.

If you fail to see *any* redeeming values in urban snow, you have to admit that, if nothing else, the big storms that deliver those humbling hard knocks to our cheeky urban ways offer guaranteed prime-time entertainment. The bigger the snowstorm, the more poignant drama, juicy suspense, and wholesome heroism will be served up by the media. "Weather is more popular than sex in this country," says Canada's foremost weather doctor, David Phillips. The badder the better. "You can hardly turn on the tube these days and not find a channel focusing on a major weather event. I call it storm porn."[177]

The next time a snowstorm comes a-calling, keep these benefits in mind. They may provide the last line of mental self-defense as your windows and doors and cherished daily routines slowly surrender to the snow. However you greet it, whether with reverence or revulsion, one thing's for sure: snow's refusal to be fenced in by arbitrary dates, forecasts, or technologies will always give our illusions of self-importance and control over nature a good run for their money.

Chapter 6
Playing in Snow

It's just so much fun you can't get enough of it.

−Miles Davis, kite-skier

My Life as a Skier

For some stupid reason known only to my foggy teenage brain, I rebelled against getting a ten-speed bike, even though all my peers were snapping them up much as hula-hoops had been twenty years before. In the early 1970s it was universally thought that the pinnacle of coolness was to be seen riding around your high school parking lot on a shiny new Raleigh or Peugeot ten-speed. It was about that time that I discovered one of my favorite words, *iconoclast:* one who assails cherished beliefs. This instantly became my chosen avocation in life, at least my teenage life—literally to be a smasher of icons of all that is cool.

Soon after I graduated and left my cool friends far behind, I quietly snuck out and bought a handsome Raleigh ten-speed and fell head over heels in love with it. It became a true freedom machine, liberating me from academic slumbers while at university. I even gave it a name, Demian, and once, when some villain snitched it, I remember wandering through the lamplit streets of Kingston, Ontario, calling out, "Demian! . . . Demain!" as genuine tears ran down my cheeks. I can't tell you how jubilant I was when I spotted Demian's half-submerged frame along the muddy shore of Lake Ontario.

Why am I opening my heart to you about all this? Because I did the same stupid thing with downhill skiing. Again, as a teenager, I became gradually disgusted by the creeping coolness that was infiltrating this noble outdoor pursuit. Perhaps I was sour-graping my own meager skills, but I swore off downhill skiing the day I saw my first pair of shin-high, pink plastic ski boots—on a *man* yet. That was about 1972. For better or worse, those pink boots and, I'll admit, sky-rocketing prices for lift tickets ended my downhill days forever. But as one chapter in my life as a skier slammed shut, another swung wide open. Cross-country skiing in those days was primarily a pastime for eccentric mountain folk and retired golfers. Definitely uncool. So naturally this is what I turned to.

I threw myself headlong into flatland skiing. For instruction I relied on a very pedestrian, sloppily bound ski manual, some of the pages of which may still be blowing around the golf course where I practiced. I quickly mastered the elusive kick step without forking over one cent for lessons. Other than a few sidesplitting adventures on telemark skis, for the next quarter-century I fastened only cross-country skis to my boots.

What about skating skis, you ask? By the time they became a rage in the early 1990s, I had developed a bum knee from overexuberant running on pothole-infested roads. For years, while I poked happily down the trail on my cross-country skis, cool kids and adults alike would streak past me on their psychedelic skaters. I watched incredulously as they whisked *up* precipitous slopes almost as fast as I could go down them. I had no doubt that similar exertions from me would elicit audible pops from my almost middle-aged knee. Still, it did look like fun. I had babied my knee over the years and the need to be, or not to be, cool no longer held sway over my life. All residues of resistance fell away when, out of the blue, a neighbor offered me a pair of half-decent skating skis for the exorbitant price of 25 dollars. That's 25 *Canadian*. I had no choice but to accept her offer and try my luck at this formerly taboo sport.

On my maiden voyage across the frozen lake behind my house, my twelve-year-old daughter Jaya—one of those cool kids who could ski-skate your pants off—announced that I looked like a duck on my new skating skis. Granted, they were a tad short for me. The woman I bought them from maybe comes up to my shoulder. But ye gad, were they fast! Like my beloved ten-speed bike—may it rest in peace—it was love at first flight. And do you know what? My knee has never been better.

Ski-skaters hone their skills under a dazzling April sun.
Tessa Macintosh

A private moment from my snow journal conveys a smidgen of the joy I have found in ski-skating:

> *The sun's glare off the snow on Back Bay is almost painfully bright. But after a long, dark subarctic winter, this is the kind of pain I can handle. Ball bearings on glass is the image relayed from the bottom of my skis. With the wind at my back and a dog at my heels, I am flying. Today I finally find the native beat of ski-skating, able to sustain it effortlessly for miles at a stretch. It is truly meditative. My concentration follows each gliding stride, intensely focused yet increasingly relaxed. Exertion and thought give way to a rhythmic repose that fills my being. By the time I round the tip of Latham Island, I am feeling quite high and confident.*
>
> *For a few moments I am in shadow, and the north wind reminds me that winter may still be lurking behind the far hills. Then I glide into the sunshine flooding straight at me down Yellowknife Bay. The rhythm stops. I am immobilized by beauty. The flaxen radiance, the generous heat from the sun, a soft fringe of mist hugging the lake's distant horizon, the shimmering of snow crystals—opal, sapphire, and onyx—the enveloping silence. The light is golder than gold. I am gripped by a sudden desire to forsake all responsibilities and reason and ski forever into this light, down the bay, and across the oceanic snowfield called Great Slave Lake.*
>
> *Instead I turn to share this moment with Haley [my dog], who I discover is about half a kilometer back. Unaccustomed to following me at such speeds, she is sitting on her haunches, giving me what appear to be dirty looks. Even at this distance her body language is clear: "Screw this noise." With a few cheery whistles, I coax her back to my side. I click once again into that magic rhythm and skate on as if in a dream. Happy. Grateful.*

As a born-again ski-skater, I now leave long chickenlike tracks all over the place. Ever faithful to my old cross-country skis, I reserve them for the deepest, raunchiest snow conditions, the thickest woods, or the heaviest packs. I have yet to find similar ecstasy in kite-skiing, but there is still hope.

Go Fly a Kite

What first grabbed me about kite-skiing was the sound, a hissing refrain sung by four piano-tight lines connecting an anonymous skier to his gently bobbing kite. I was merrily ski-skating across the lake behind my house one bright April day, smugly convinced that I was pushing the envelope of human-powered speed over snow, when I had to stop and listen to that sound.

A goggled man wearing wide telemark skis was standing stock-still, squinting up at his ruby-red kite with great concentration, as if engaged in a silent countdown before launching into space. About the size of a long oval coffee table, the kite seemed to pulse with latent power in the brisk spring breezes. I leaned forward on my ski poles, spellbound by that ethereal hiss, the waggling kite, and the blinding sun. I watched the man rotate his skis to some apparently magic angle, give his kite a subtle yank, then take off like a racehorse across the dry packed snow.

I later learned that this man, who we can now call Chris, always carries a global positioning system (GPS) with him while he's flying along beneath his kite. Among the nine hundred or so setting options on a typical GPS (I'm still working on latitude and longitude) you can rig these things to capture your fastest speed. The snow on Yellowknife Bay is particularly attractive to kite-skiers, being dry and shallow and almost uniformly flattened by countless passes of virile teenage males going nowhere fast on overpowered snowmobiles. After a particularly bracing run down the bay, Chris once logged on to his GPS to discover that he had hit a top speed of 67 kilometers (42 miles) per hour. I understand that he is gunning for 75 kilometers (47 miles) per hour next winter.

"These things generate so much power," says Miles Davis, a passionate convert to kite-skiing. "The thing about kites is that there is almost nothing to them. They roll up into this tiny little ball and weigh next to nothing. There are no hard moving parts, nothing to break. It's just fabric and string pulling you along. Yet they can get you screaming across the snow with power to spare."

Since you are probably going to ask me, I'll tell you now that Miles Davis was indeed christened after the celebrated trumpet player of the same name. His Scottish father is an avid jazz musician who once met the famous musician, collecting and framing an autograph for his young son that read, "From Miles to Miles."

Though Davis has been kite-skiing for less than a year, like many recent converts—whether in religion or sports—he speaks of it with an almost evangelical zeal. "It's an irresistible sport," he told me in his cheerful Scottish brogue. "Everybody that's taken it up begins by just standing around watching us. Then they come over and try it. Before you know it, they're hooked. If you like downhill skiing, it's got the same adrenalin rush without all the costly lift fees and infrastructure that go with it."

A kite-skier breaks a speed barrier while sailing across Great Slave Lake. Devotees say this burgeoning snow sport provides the same adrenalin rush as downhill ski-ing, but without the costs.

Miles Davis

Our local conditions are so ripe for kite-skiing that European adventure-seekers train here en route to the North Pole *and* the South Pole. As I write (January 2003), two ardent Irishmen that trained here are on a 650-mile (1,040 km) kite-skiing extravaganza from the South Pole to Patriot Hills at the edge of Antarctica. A few years back, some of these adventurers-in-training sold off their equipment before flying back to warmer climes. That's how the sport took off in our neck of the woods.

Kite-skiers have become such a common sight out my back door that a neighboring five-year-old, while being outfitted for her very first cross-country ski lesson, looked up at her instructor and asked, in all sincerity, "But where's my kite?"

Davis stepped out his back door the other day and, with Chris and his wife Jennifer, embarked on a nine-day, 200-kilometer (125-mile) kite-skiing expedition across the fifth largest lake in North America. "The idea was to ski across Great Slave Lake to Hay River. We're all sailors and canoeists in love with the concept of unpowered travel. We were really curious to see if we could use kites as another form of transportation."

Soon after setting out, the wind died. They had no choice but to start man-hauling their gear across the snow.

"We had two converted snowmobile sleds," said Davis. "Mine was pretty heavy, about 150 pounds. It was so exhausting. I remember sitting out there with all that gear thinking, 'Okay, now, what can I remove?' The only thing I could come up with was one spare pair of underwear. Everything else was absolutely essential." (I never did get where he stashed the underwear.)

The wind picked up the next day, and under kite power alone the expedition clicked off almost 50 kilometers (30 miles). It wasn't long before they completely lost sight of land. "The further we got from shore, the more snow had been blown clear by the wind. We'd stop, totally out of sight of land, and look down through perfectly smooth black ice at the water just a few feet away. It was a little creepy but very beautiful."

Farther on, they discovered that too much wind can be as taxing as too little. Locked into gusty winds with a heavy load on his tail, Davis's knees gradually began to cave in. "There was one point where I was totally exhausted and could barely stand up. I swapped my sled for Jennifer's smaller kite. There was always one person not pulling a sled so we could carry on without somebody keeling over."

Things got hairier when the expedition ran into a jumbled field of early winter lake ice that had been crumpled and thrust upward through the snow. "It was very fast out there," recalls Davis. "The sunlit chunks of broken ice were gorgeous to look at, but deadly sharp. It was pretty scary, like going blind through a minefield, as we whipped past them at around 50 kilometers an hour."

They woke the next morning to perfectly clear skies, mild temperatures, and not a puff of wind. "So we started man-hauling again. It was so slow." Same story on the fourth day. With more than 100 kilometers (60 miles) of snow and ice between them and Hay River, the kite-skiers realized they weren't going to make it.

But, for Davis, the experiment was well worth it. "Maybe we should have started with a shorter test drive. But the beauty of this new technology is that you can push the edge without taking huge risks. Once we decided to abort our plan, we just pulled out the satellite phone and within 40 minutes there was a ski-plane at our feet. It was disappointing on one hand but very satisfying on the other. We flew home at 15 meters off the deck, following the exact course we'd kite-skied. The snow-covered ice just flew past us at 160 kilometers an hour. It was a good feeling, knowing we'd skied that distance powered mostly by the wind."

Davis's latest kick is kite-jumping. A big ice road goes down the middle of Yellowknife Bay, the plow margins of which feed giant snowdrifts. Prime jumping time is early spring when the ice road closes and the drifts are tallest. "You come screaming straight for them, yank the kite up, it lifts you off the snow, and you literally fly across the road." At the tail end of the season, when the rotting snow is pockmarked with puddles and occasional open leads, kite-jumpers use their aerial skills to lift themselves up and over the water. Another skier told me how he'll shoot over a stretch of water "without even making a bow wave."

All winter long, Davis and his kite-skiing pals keep a close eye on the wind through a government weather webpage, bookmarked on all their computers.

"In the winter we're always watching that. The moment the wind looks good, we fire the news around. 'To hell with work. Let's get down to the ice.' That's the thing about being a totally weather dependent sport. You can't just wait around for the weekend. Quite often we take a two-hour lunch just so we can get our fix of wind."

I asked him how this went over with his employers at the local museum.

"Oh, they don't mind as long as we put in our hours. We're much happier and more productive after a good session of kite-skiing. It's so nice to have the wind, and nothing but the wind, pulling you along. The dazzling color of your kite against the sky and the snow is fantastic. It's just so much fun you can't get enough of it."

On Velvet Shoes

"In the beginning," goes a Chipewyan tale from northern Saskatchewan, "man did not exist.

"Then suddenly there was man, it is said. Who made man? We do not know. When winter came, man made himself some snowshoes." This did not come quickly or easily to man, having little, if any woodworking experience.

"How am I going to do it?" he wondered. Guided mostly by the seat of his caribou-skin pants, he cut down a birch tree, managed to build a passable frame, installed the crossbars, then brought the project to a halt.

"How will I do the latticework?" he thought, for he had no woman to do it for him—as was the custom in those earliest of days. He lay down for a long winter's nap, and when he awoke he saw that his snowshoe frames had been partially laced.

"Who could have laced my snowshoes while I was asleep?" he said to himself, for he was quite alone. This construction by magical increments went on for several nights, to the mounting puzzlement of man. Then one morning, just as he was waking up, he saw a ptarmigan flying up through the hole in the roof of his tipi.

"Ah-hah! It must have been the ptarmigan!" he exclaimed.

Again the next night he went to sleep and at daybreak, through shifty half-closed eyes, he glimpsed the ptarmigan flying up and away. He looked at his snowshoes. They were almost completely laced.

"Hmmm. I know just what to do," he said to himself. That evening he covered the top of his tipi with the skin of a black bear. Then he lay down to sleep, smiling to himself.

When man woke the next morning, his finished snowshoes were lying beside him. There, in the shadows, was the ptarmigan. It looked cautiously at man, startled to see him awake so early. "I must fly away quickly," she thought, but when she fluttered up to the tent roof she discovered that her escape route was blocked. She dropped in defeat to the tipi floor and turned instantly into a beautiful woman with long silky hair. As the story goes, "the man and the woman slept together, of course, and in time they had many children. We are their descendants."[178]

The exact shape and size of those primeval snowshoes is left a mystery. No doubt ptarmigan-woman had the local snow conditions in mind when she designed them. Most of the traits peculiar to traditional snowshoe styles used by native people across northern North America have everything to do with variations in snow. The shape and angle of the toe, the tightness and pattern of the lace-work, the number of crossbars, the presence or absence of a toe-hole, the width and length, the total surface area of the frame—all these features vary from region to region in concert with the prevailing density, depth, weight, and moisture content of snow.

For instance, in the central subarctic of the Northwest Territories, where some of the continent's lightest, driest, and fluffiest snow falls, traditional Dene snowshoes are typically long and narrow for maximum stability and speed. They have sharply upturned toes, making it next to impossible to catch the tips in deep snow. They are laced with a tight hexagonal weave for maximum flotation. To the east, where the snow is deeper and heavier, the Cree favor the huge round "beavertail" design that can measure 0.5 meters (1.5 feet) in diameter. It incorporates a very fine babiche mesh into which the elders often weave images of guardian spirits to boost hunting success on the trail. "These shoes," writes one avid snow walker, "are of such stunning quality and artistry that nothing could be added or subtracted without jeopardising perfection."[179]

And to the west from Alaska come perhaps the biggest snowshoes ever made, the so-called whales, designed for the deep mushy snow found on the southern slopes of the Alaska Range. These monsters include several crossbars, sharply upturned toes, and a tip-to-tail length of well over 2 meters (6.5 feet). Says one devoted whale user, "They're not fast, but boy do they float!"[180]

Snowshoeing is a very unglamorous business compared to high-profile snow sports like heli-skiing or snowboarding. If you want to snowshoe without sacrificing your coveted sense of flair, you can always opt for a high-tech, high-cost pair of snowshoes fashioned with NASA-approved neoprene and titanium alloy frames. But once you get out there, who's really looking at your shiny gear? By definition, snowshoeing is largely a backcountry activity where practicality and simplicity rule.

*Snowshoers revel in the far backwoods. Whether your snowshoes are made from
titanium or tamarack, if they float, they'll get you out there.*
Tessa Macintosh

To get started, what you really need more than anything is a zest for exploring the bush. For about a hundred bucks, any decent pair of snowshoes can take you places you could never get to by any other mode of travel—and I include here snowmobiles, bulldozers, and helicopters. The only thing that might rival the snowshoer's maneuverability in a wild rolling forest is a jetpack. To get there, you don't need a rigorous training schedule. The only prerequisite is that you are able to walk. Sure, at first there's a bit of a knack to not stepping on your snowshoes or snagging your toes in the harness. What works for me is imagining myself rambling along with a semi-syphilitic gait. I've seen three-year-olds catch on in minutes, without the crutch of such mildly vexing visualizations.

The soothing beat of each step, the friendly crunch and squeak of snow underfoot, the sudden encounter with a rabbit or deer scampering across your path, the smug knowledge that, at a sustained easy trot, you are burning upward of a thousand calories an hour—these are among the wages earned by day and by night to deposit in the bank of a snowshoer's heart.

One balmy New Year's Eve—and by "balmy," I mean anything warmer than −20°C (−4°F)—my family and I stepped out from the merry glow of our River Lake cabin and strapped on our snowshoes for a full-moon jaunt. I'd like to see you try this kind of thing on a snowboard.

A snowshoer's view of a snow-filled forest aglow under the northern lights.
Tessa Macintosh

The only color I can think of to describe the snow-filled, moon-washed forest is that blue-green sheen of the Caribbean Sea. This is a poor approximation of the resplendent glow that bathed the poplar woods behind our cabin. We set out, a snowshoed family, under that beautiful light—no, in that light. All four of us were determined to shun anything resembling a packed trail, opting instead for the untouched blue-green powder that twinkled—yes, twinkled—under the startling moonlight.

Things really got twinkling when, after a clamorous countdown to midnight, we launched an illicit emergency flare over the lake, cheering in the New Year with a huge flash of blood-red snow. On the way back to the cabin, we zigged and zagged through the dense woods, discovering open ribbons and islands of perfect snow walled in by adolescent spruce, all frosted and still.

In *Wintergreen: Reflections from Loon Lake,* Monte Hummel, president of World Wildlife Fund Canada, reveals one of the best ways to shake off the tentacles of the urban workplace: head for the bush, don your snowshoes, and go howling for wolves and coyotes under a full moon. While en route to nowhere in particular behind his cabin in backwoods Ontario, he stops on a high ridge "long enough to allow silence to wash out the rhythmic sound of my snowshoes breaking trail."

> By far the nicest time to travel at night is in midwinter, using moonlight and snowshoes to get around. On a bright night, tiny snow-diamonds show themselves scattered between moon shadows. But even without a moon, snow reflects a surprising amount of soft night light, illuminating a natural night world every bit as wonderful as our more accustomed day-lit universe.[181]

After a lively conversation with a coyote, Hummel stops again near a beaver pond to build a quick fire and make some snowmelt tea and bannock. Here he pays a final tribute to the value of snowshoes even while not attached to his feet. "Tea and bannock are best enjoyed together, outdoors under the stars by a fire, sitting on one snowshoe laid flat as a seat, and leaning up against the other jabbed tail-first into the snow as a chairback. There are really no other choices worth mentioning."

The one and only drawback to snowshoeing at night is that you can't turn around and fully drink in the ineffable beauty of the trail you've just made. I thought I was alone in experiencing the gratifying rush of this sight until I picked up a copy of *A Snow Walker's Companion,* by Garret and Alexandra Conover. In this snowshoer's bible, they quote a 1930s bushman who did his best to capture the ineffable in words.

> A snowshoe trail on a sunny day after a light fall of snow is a lovelier thing than I can describe. I often look back at it stream-ing from our heels, flowing astern. . . . A darker serpentine rib-bon, scallop-edged, filled with tumbled blue shadow markings. And every individual print is a beautiful thing. It is like sculpture and like painting, endless impressions of an Indian craftsman's masterpiece.[182]

Hard-core snowshoers seem to take great pride, not only in the sculp-tures trailing out from behind them, but also in the palette of sounds drift-ing up from their shoes. Owen Neill is a wolf researcher and poet from northwestern Ontario. His poem "Breaking Trail" tells me he has spent a lot of time harkening to the music of his snowshoes.

> *There is something freshly alive*
> *about being first*
> *to break trail*
> *across a field*
> *of new snow falling.*
> *Bright whiteness*
> *pulls the eye,*
> *transforms a field*
> *into an ocean*
> *to be crossed*
> *particularly*
> *in a winter twilight*
> *when no shadow falls*
> *to add dimension*
> *to a flat glowing world*
> *where feet are prows*
> *of winter ships*
> *on a winter sea*
> *creaking, crunching*

as the mood sees fit to follow
leaving a wake
of foot furrows
in this scape of floating flakes.
And the rhythm of this walking
backs a kind of symphony.
Bass of leather
rumbling with the tenor squeal
of hard-packed snow
against rolling mid-range soles
with overtones galore,
make this walking music
joyful as carols
by friendly fires.
On and on as a ship roams
eyes seek horizons
through rainbow tears.
Ears revel
in this pristine concert hall.
Mind loses all bounds
yet balances somehow
in a delirium
of diamond ecstasy
as our manship plows on
beyond imagination
in our own untraveled worlds
to rival even old Ulysses.[183]

Slip Sliding Away

There's a good reason why ski trails get names like Pipe Dream, Cloud Nine, Utter Abandon, and Ecstasy. That exhilaration, that freedom, that joy of flying over snow is basically the same whether you're on skis, a toboggan, snowboard, bobsled, kicksled, or (and this is the latest gear for snow play that I've seen) a Snowfer, winter cousin to the windsurfer. The same rush would apply if you happen to be competing in the annual furniture race at Colorado's Big Mountain ski resort and are screaming down a hill on a ski-mounted bed frame, toilet, coffin, bathtub, or a junked Harley-Davidson motorcycle frame.

Snowboarder finds ecstasy in flight.
Tessa Macintosh

"Snow arouses playfulness," says Bernard Mergen in *Snow in America*. "Snow, like play, is extraordinary, a break from everyday routines...." But, he adds, "It is as impossible to define *snow* as it is to define *play.*"[184] There is, however, a common denominator to all this. What makes snow so much fun is what's going on underfoot, no matter how you're dashing through it.

As the surfer pines for the ultimate curl and the sailor for perfect winds, the slider of snow craves for that magic balance between frictionless flight and surefooted control. More than one skier-poet, having found that groove, has coined verse that Mergen calls "practically orgasmic." Take, for instance, Walter Eaton's poem "Skis."

> One last deep breath of stinging air, and slipped
> My skis across the rim: then farewell breath,
> And almost vision, too, as tears rolled down
> My cheeks, while past my face the riven air
> Tore by, and all the hillside flew to meet
> My flying figure with a low-hissed song—
> The song of rapid runners cleaving snow![185]

Mergen follows this vein by citing an anonymous poet obviously steamed by his fling in the snow.

Straight run our skis down the untracked steep,
Seaming a virgin face.[186]

When it's good, it's good; when it's not, it's not. There are a lot of forces working against you if you're lusting for that perfect plunge down the mountain. The only sure thing working in your favor is gravity. One competitive skier's manual tells me that if you've got the right "tuck" and are dropping fast enough (anything over 100 kilometers [62 miles] per hour, apparently), you can also put aerodynamic lift on the plus side of this equation—if you dare. All other forces are hard at work trying to stop you in your tracks.[187]

At the *macro* level, with the wind in your face and your skis or snowboard or whatever skimming over the snow, the forces of air resistance, plowing, shearing, and compression are opposing your fall through space. At this level the general density and texture of snow dictate the impact of these forces on your descent. The scenarios are as varied as the snows. Wind slabs and crusts will throw the most curves at you. You may be sailing merrily along on the surface, welcoming a wind-hardened crust, when suddenly you plunge through, only to vault out, then plunge in again, this time planting your face in the snow for artistic effect. The snow pros call this "erratic penetration."

On the other hand, a tough even crust may be so hard that your edges can't bite, and you end up sprawling over the snow like a wave-tossed starfish. Ripples and ridges, cracks and cornices, sunballs and sastrugi—these are among snow's palette of "surface irregularities" that stand ready to thwart your downward path. A keen eye, the right fall line, and a bit of luck can help you work with these features and forces instead of against them—that is, unless they've been stomped out of existence by snow-grooming machines (which I'll get to in a moment).

At the *micro* level, where the pedal hits the metal, so to speak, things get a lot more cryptic and hence more difficult to control. At this scale, friction holds sway directly underfoot, exerting its influence through subtle forces not fully grasped by science. To explain "fast" snow, early researchers studying the physics of skiing jumped on the "ball-bearing" hypothesis, speculating that the skier was ferried along by billions of round, free-rolling snow grains. Though on a good day, it sure may *feel* like ball bearings down there, most snow grains, no matter how old or rotten, rarely occur as disconnected spheres.

Plunging through the powder. When all you've got working in your favor is gravity, it's what's underfoot that counts.
Northern News Service

Falling
for
Snow

As a fledgling skier, the story I grew up with was that as I skimmed along, the pressure of my skis generated sufficient heat to create a thin film of slippery meltwater. Taking this process to extremes, some glaciologists believe that a heavy glacier on the move is actually sliding on a microfilm of molten rock. Unless you've had an exceptionally large lunch or are traveling close to the speed of sound, you can forget about melting rocks along your path. But snow? Yes. Field experiments in both the United States and Japan have confirmed that the pressures and speeds associated with skiing or other downhill sports do indeed create a "frictionally melted water film" under-foot. Apparently my boyhood notion that I was actually waterskiing on snow was accurate. What has not been confirmed, however, is how slippery this film really is. Everybody knows that friction on snow is least when the thermometer hovers right around the melting point. Hence the bliss of spring skiing, for instance—and the agony of squeaking over bone-dry snow at −40°C (−40°F). The traditional explanation is that at temperatures close to 0°C (32°F), the meltwater film is at its slickest. So the uninitiated might think: the wetter the better. But why then is movement over snow such an absolute drag at warmer temperatures when the snow is virtually awash with meltwater? Nobody knows for sure.

Okay, so maybe it's simply the intrinsic low friction of ice that lets us sail over snow when the temperature is right. Scientists in this camp view ice as a "solid lubricant," analogous to graphite. What lets you push your graphite pencil around with such ease is a molecular structure of stacked planes that, when rubbed, can slide easily over one another then flip back into place. A 1955 study at the University of Minnesota ran with this idea, modeling snow "asperity" (surface roughness) features as "elastic-plastic microscopic protrusions." After filling several chalkboards with scribbled equations, researchers concluded that the icy surface of snow grains is coated with a "liquid-*like*" layer several molecules thick that helps take the bite out of any protrusions when squashed by your flying feet.[188]

Twenty-five years later, a think tank of Colorado physicists was still burning up the chalk, trying to get their heads around the unusual frictional qualities of snow, particularly near the melting point. They posed arcane theoretical arguments, conjuring layers upon subatomic layers of "mobile protons, disordered atoms, and larger scale defects" that collectively smooth out the microbumps in snow. They came up with some great lingo to explain all this—plastic deformations, asperity fractures, trans-lubricating films—but in the end all they could conclude with certainty was that "the surface of ice is unique compared with any other materials." Another researcher, stymied by a competitive skier's uncanny success at picking just the right combination of waxes, conceded that the physics of gliding over snow is "virtually unknown."[189]

Japanese tourists brave subarctic slopes on a sealskin. The nitty-gritty physics of gliding over snow—no matter what you're riding—remains a mystery.

Tessa Macintosh

*The going gets easier for these dogs as new frontiers are crossed in the technology
of reducing friction over snow. Today's Teflon sled runners may soon be replaced by
something even faster.*
Fran Hurcomb

Not knowing exactly what's going on at the snow surface hasn't stopped commercial snow barons from developing elaborate technologies to enhance the downhill sportster's speed and control. Grooming the snow-clad slopes (as one might a putting green) became the order of the day soon after skiing took off as a winter sport in North America during the late 1930s. Over the next few years all manner of machines took to the hills. Among the menagerie of tractor-, donkey-, and human-pulled devices was a roller built in 1952 at Winter Park, Colorado, slapped together from rolls of electrical tubing, screen door springs, aircraft cables, and bicycle wheel rims. Fifty years later, tanklike Beartracs, Sno-Cats, and light-duty dozers now roam the downhill Meccas compacting, smoothing, chopping, tilling, and harrowing the snow like so much arable soil. The promised harvest in this case is a homogenized snow surface engineered for gravity-powered flight.

Like corn and potatoes, the snow groomer's harvest does not always turn out as expected. Among the seventy-odd terms used to describe commercialized snow are "death cookies" (nasty chips frozen to the snow surface, often left behind by grooming machines) and "kitty litter" (rough debris caused by shoddy Sno-Cat work).[190]

When the snow runs low on the slopes, as it often seems to these days, you simply make more. Experiments in snowmaking began in the late 1940s when Mohawk Mountain in Massachusetts became the target of irrigation equipment borrowed from surrounding valley farms. After a lot of trial and error, including a few skiers' broken limbs and lawsuits, the technology crossed the threshold into making bona fide snow—as opposed to elaborate slush—in which the cooling effect of pressurized air converts an atomized water stream into crystals of ice. By tinkering with the ratio of water to air, snowmaking machines have evolved to the point where different types of snow can be manufactured on demand—from dry powdery snow to a fine glaze of ice. If the temperature is not cooperating, fear not. You can always stir some Snomax into the mix, an amazing fluid created by Eastman Kodak laboratories that contains bits and pieces of bacteria that serve as nuclei for the growth of snow crystals. The product label promises to "produce more snowflakes, at higher temperatures and with less damage to the environment."[191] Designer snow may soon fall within technology's grasp.

Whether slipping across a manicured bunny hill or schussing down an untrodden summit, you've got to keep at least one eye on the snow. As the surfer reads waves and the sailor reads wind, the slider of snow translates its shape, texture, and sheen in terms of speed, grip, and glide. Like any language, snow literacy requires that you use more than your eyes to get the full message.

Knowing how to read ripples on a snowdrift or the curve of a cornice is not enough for Edward LaChapelle, a longtime snow-watcher based at the University of Washington. The guy should know. Among his stellar list of snow titles are *Field Guide to Snow Crystals, ABC of Avalanche Safety,* and his latest, *Secrets of the Snow: Visual Clues to Avalanche and Ski Conditions.* To maximize our downhill pleasure and safety in snow country, he advises you to focus on much more than just visual or atmospheric clues when interpreting snow conditions. The sudden *whumpf* of settling snow tells you to "Clear out! Avalanche risk is high." A particular rattle of wind-driven snow against your parka hood speaks of different crystal shapes that may, or may not, speed your drop down the mountain. The kinesthetic sense of changing snow layers beneath your feet tells how soon you might go ass over teakettle if you don't hurry up or slow down. If you truly want to call yourself a snow literate, Professor LaChapelle insists that you bone up on these more subtle signals as well. He admits, though, that this takes a certain knack. "[These signs] form just as important a part of interpreting snow conditions as the thickness or density of a new snow layer," he writes, "but are relegated to the realm of seat-of-the-pants feel or intuition."[192]

The upshot here is that, if you're out there playing in the snow, the best way to learn its language is by attending to its many moods and messages *with all six senses.* Like Chinese or Russian, the language of snow is by no means easy to pick up on the fly. Ask a seasoned Inuit seal hunter, for instance. Even high-paid snow scientists, hired by downhill resorts or sporting organizations to squeeze more speed from the snow, often find themselves stuttering. Could it be because they are dealing with an always unpredictable, constantly changing, frankly mysterious substance whose attributes are, in a word, kaleidoscopic?

Ironically, all it takes is a seat-of-the-pants assessment to identify the Cadillac of downhill snow conditions. You don't need a Ph.D. in glaciology to spot it. And keep your Sno-Cats and snowmakers and bacteriological spray guns well back. Like all the best things in life, technology simply cannot improve on this. I'm talking here about powder. Deep, dry ecstasy inducing powder.

In *The Guinness Book of Skiing,* British Olympic skier Peter Lunn describes what is to him—and he's not alone on this—the summit of playful pleasure in the snow: seaming the "virgin powder" on skis. Leaping quickly from mechanics to aesthetics, Lunn explains, once and for all, why the best is the best.

> Powder snow is the best of all skiing surfaces. With correct technique, it presents little resistance to the turning ski, and yet provides a soft resilient surface against which the ski banks as it comes round. The nonskier, who has waded through powder snow, sinking at every step, may be surprised at my use of the word resilient. . . . But powder snow is like water; it *is* resilient, not for the static, but only for those in motion. So long as the skier keeps his skis pointed upwards, his tips will rise to the surface of the snow as he gathers speed. And, when he turns at speed, the snow will provide a smooth elastic support against centrifugal force. On hard snow the skier must struggle to maintain precision. But as he turns on powder, the relationship between him and the element has the gentleness of a caress.

Indeed, the Austrians have a saying that there are in life two supreme sensations, and one of them is skiing in powder snow.[193]

I can't honestly say whether you would find the same kind of ecstasy cleaving the powder with a ski-mounted bed frame or bathtub. Perhaps with practice. In search of similar tingles, an increasing number of offbeat snowboarders are taking to the untrammeled powder. Snowfers may be next. As snow-sliding technology evolves over our lifetimes and beyond, who knows what kind of contraptions will be streaking down the slopes and who'll be riding them?

But I can tell you one thing for sure. Whatever they ride, they'll all be hankering after the same sweet thing: that rousing joy that only a good plunge through the snow can offer.

Epilogue
Seeing Snow

Try to describe snow, and immediately there arises a context.

—Ruth Kirk, *Snow*[194]

Only once have I come close to being carsick. I was trying to knit a scarf for my kid's teddy bear in the back seat of a car winding up a hairpin-studded road in British Columbia. As a vagrant teenager ferrying across the English Channel in a moderate gale, I discovered that I don't get seasick. Conducting wildlife surveys for mountain sheep in the Yukon, while our plane spun circles so tight it seemed to hang on one wing, I realized that I don't get airsick either. But none of this could prepare me for my journey across the frozen sea in a bug. The time, many years ago, when I got bugsick.

Well, it wasn't really a sea. It was Hudson Bay. And technically it wasn't a bug, though that's what everyone calls them. It was a Bombardier (pronounced *BOM-ba-deer),* one of those big beetle-shaped snow crawlers with tank tracks and skis. Some northerners use them as mobile ice-fishing shacks. "The guy who invented these things really knew what he was doing," says Vern Steinwand, a fisherman and loyal bug owner from Hay River, Northwest Territories. "There's never been anything built quite like 'em."[195]

Most of the bugs still in action—and there's a surprising number of them where I live—were built over fifty years ago. Though virtually indestructible, a good suspension was never one of their strong points.

Hay River fisherman Verne Steinwand shows off his handyman's special: a fifty-year-old Bombardier bug, ready to roll.
Northern News Service

The one I bumped around in was a taxi. My driver, George (whose Inuktitut last name went right over my head), was delivering me from Arviat to Whale Cove, Nunavut, on the storm-ridden western coast of Hudson Bay—from one blip of civilization to another. The combined population of both hamlets fell below fifteen hundred residents, 95 percent of which were Inuit. My bold trek was inspired by some earthshakingly important government business, the nature of which completely slips my mind.

We left Arviat late in the afternoon, which in an arctic mid-January comes soon after lunch—something I later regretted eating. For most of the 150-kilometer (90-mile) journey, it was pitch black outside. All of it involved travel over snow-caked sea·ice. Only one headlight was working and that at a cock-eyed angle. The whole way, it illuminated nothing but dolphin-sized snowdrifts which, from where I sat, had the consistency of concrete. My berth was an oily wooden bench under a wall of chains that rattled, not far from my head, like a parade of Christmas ghosts.

When we stopped for a pee break, I asked George about the chains.

"They're to pull us out if we go under," he said matter-of-factly.

I was too green to ask him to elaborate.

Besides the fact that I survived, the most amazing thing about that journey was how George ever found Whale Cove. He didn't talk much. He was busy looking at . . . what? No one had built a queue of Inukshuks (traditional stone figures, often used for navigation) to guide the odd bug through this trackless terrain. Imagine the labor, hauling rocks out there every winter, building a strung-out platoon of Inukshuks, only to have them sink to the bottom of the bay every spring. I peered out through one of the bug's tiny round windows, looking for stars, thinking, *Maybe that's how he did it.* But no. Overcast. Not one twinkle. I hadn't a clue how he navigated.

I was green back then in more ways than one. As a recent immigrant laborer from the south, I assumed that George drew on some obscure Inuit instinct which I would never understand. In the meantime, I trusted him with my life. When the lights of Whale Cove suddenly pierced the darkness ahead, my curiosity became uncorked. This navigational feat was sheer magic to me, and I simply had to know his secret.

"How the *hell* did you do that, George?"

"Lucky, I guess," he said, with a faint shrug.

Polite grilling of more-talkative residents in Whale Cove helped me solve the mystery of that magical (though nauseating) tour across the bay. "Drifts," one man said. "It's all in the snowdrifts." To me, those drifts were the bane of our trip. To George, they were a boon. He must have had his eye set on them the whole way. And not just on any drifts but specifically the ones with a high tongue-shaped bump on the upwind side. These, I later learned, are what George's people call *uqalurait,* from the word *uqaq,* meaning "tongue." I was told about other kinds of drifts and squiggles in the snow, but it became clear that it is the *uqalurait,* formed by the prevailing northwest wind, that best serves the Inuit navigator.

Other winds form tonguelike drifts, but none are so distinctively shaped, so hard, or so persistent. In his book *The Arctic Sky,* Igloolik historian John MacDonald reports that, if in doubt as to the directional origin of a drift, Inuit elders will test it with a snow knife. It takes some muscle to poke through "the good ones." Even after a jumble of winds or a fresh, smoothing snowfall, all they need to do is kick away the new snow to reveal the guiding form of *uqalurait* below.[196]

The rock solid reliability of *uqalurait* drifts is an artifact of the wind that shapes them. The prevailing northwesterlies, called *Uangnaq,* are strong, gusty, prolonged, and often bitterly cold. These forces combine to create the distinct qualities of *uqalurait*. Says Igloolik elder Theo Ikummaq, "It dies and then it blows, dies and blows, and that is why it forms *uqalurait*."[197]

What do you see in this jumble of arctic snowdrifts—a road map, a compass, a sore backside?

Tessa Macintosh

The apparent fickleness of this wind earns it a feminine face for many Inuit. On the other hand, the southeasterly *Nigiq* winds tend to be milder, steadier, and shorter, creating weaker drifts, if any at all. "When it blows it is usually constant because this wind is a man," says Ikummaq. Not surprisingly, the to-ing and fro-ing of these contrary winds is often likened to the perpetual state of creative tension between partners of the opposite sex.

The Inuit have names for snowdrifts formed by winds from just about every point in the compass. Conventional scientific wisdom would lump most of these under the term "sastrugi." One glossary of Inuktitut snow and ice words contains over a hundred terms that describe subtle differences in snow, which most southerners wouldn't notice if they tripped over them. Inuktitut words account for, among other things, the origin, texture, shape, density, age, utility, and moisture content of snow. Besides all manner of snowdrifts, the Inuit have a word for the best kind of snow to chink cracks in drafty igloos: *qikuutitsajaq*. They have a word for "small pillarlike protrusions of snow formed after the soft snow around animal tracks has been eroded by a blizzard": *qaqiqsurniq*. So fine is the language's grasp of snow that it is labeled by the sound it makes underfoot. If the snow makes a loud squeak or crunch, you know you're treading on *qiqiqralijarnatuq*.[198]

The Inuit fondness for snow words has become a popular notion far beyond their Nunavut homeland. Over the past 150 years, non-Inuit visitors have jacked up the number of snow words attributed to the Inuit from four, in John Washington's 1850 best-seller, *Eskimaux and English Vocabulary for the Use of Arctic Expeditions,* to "200 or some other large number," according to ecologist Elliot Norse in 1990. One writer describing the crab industry off the eastern US coast says, "As [the Inuit] have many words for *snow,* and Arab Bedouins for *camels,* so do Smith Islanders for *soft crabs.*" A reporter commenting on the proliferation of crime remarks, "Violence to the American has become as varied and omnipresent as snow to the Inuit."[199] Canadian novelist Margaret Atwood popularized the notion that the Inuit have 48 words for snow—or was it 52?

Anthropologist Laura Martin concludes that "the structure of [Inuktitut and its many dialects] means that the number of 'words' for snow is literally incalculable."[200] Meanwhile, linguist Geoffrey Pullum says that, "In the study of language, one case [of exaggerated beliefs] surpasses all others in its degree of ubiquity: it is the notion that Eskimos have bucketloads of different words for snow." He concludes that this question has been muddied by "scholarly sloppiness" and "amateur embellishments."[201]

Whether understated or overstated, I'm content to let the "snow-word hoax" debate continue without me. Frankly, my Hindi is better than my Inuktitut. If pressed on this issue, I'd throw my cards in with arctic anthropologist Ellen Bielawski, who claims that, "Inuit knowledge resides less in what Inuit say than how they say it and what they do."[202]

Saying it is one thing; *seeing* it is another. In the unlikely event that I could ever communicate with my bug driver George, in his own language, that doesn't necessarily mean I could then see all the snowy nuances that he does. This is where instinct and ancestry govern perceptions of snow. While arctic linguists and anthropologists continue to hurl verbal snowballs at each other, the fact remains that the Inuk's unique vocabulary of snow words reflects an intimate and practical knowledge of snow developed over thousands of years. Snow remains an indispensable thread in the fabric of Inuit culture and identity—quenching thirst, offering shelter, improving access to the land, and providing a trusted navigational key that unlocks their world and makes it their own.

Wearing traditional caribou-skin pants, an Inuk near Holman Island, Northwest Territories, musters his dogs for a mush. Inuit knowledge of snow resides less in what the people say than in what they do.
Tessa Macintosh

But even to the Inuit masters, snow can also be a mixed blessing. I feel a cultural drawbridge go down when I sense that Igloolik elder Abraham Ulayuruluk has shared the same misgivings that I once did when confronted by a sea of bone-jarring drifts on a long trip across the snow. "They are not what we want," he says in *The Arctic Sky*. "They are nothing but a hindrance on our journey, but they have a use that we surely cannot do without."[203]

Soon after my stomach settled down from that bug trip with George, it dawned on me quite viscerally that the longer snow is in a people's face, the deeper is its imprint on their hearts, minds, and culture (ask the Arab Bedouins about sand). In the high arctic kingdom of the Inuit, snow steals the show for most of the year. The supreme degree to which they have adapted to snow shows just how far we humans can go in appreciating its many faces and releasing its astonishing potential. Could there be a finer expression of this intimate bond than the igloo? I've yet to find one.

Arctic scholar John Moss describes the igloo as "the nearly perfect adaptation of landscape to human needs." Calling it a *snowhouse* is a flagrant cultural faux pas, he says, since the concept of house "has no meaning to a people for whom the land itself is home."[204] A couple of skilled Inuit can build a family-sized model in under an hour. A CNN reporter recently claimed to have discovered the world champion igloo builder living in—where else?—the Nunavut hamlet of Igloolik, meaning "place of many igloos." This master builder is said to be able to whip together a sturdy two-person igloo in under 15 minutes. The apparent ease of Inuit snow masonry belies the igloo's exacting structural requirements and the care taken to select just the right snow—and I do mean *the* right snow.

In my ramblings through the written world of snow, I found several entertaining testimonies, both historic and contemporary, of non-Inuit visitors, or *Kabloonah,* trying their luck at building an igloo. My favorite story, in *North to the Nigh: A Spiritual Odyssey in the Arctic* by Alvah Simon, tells of a polar adventure in 1994, when Simon's 36-foot sailboat became trapped in the ice, then buried in drifting snow.

Looking for some much-needed elbow room in his claustrophobic quarters, he resolved one day to build an igloo. He started by sawing blocks from the handiest snow around—right off the tilted bow of his boat—but these invariably crumbled or split. After going below "to consult the Inuit masters" in the pages of a book on arctic exploration, he was soon back outside, this time far from his boat, walking back and forth over deep snowbanks while *listening* to the snow. The translated words of an Inuit elder guided him to "the perfect building material." Simon knew it the moment he heard it: that same "tortured screech produced by walking on Styrofoam." Then and there he fell to his knees and, with a rough-toothed saw, started cutting.

Arctic snow, the perfect building material—plentiful, easily shaped, surprisingly strong, incredibly light. Perfect, that is, if you know what to do with it.
Lorne Schollar

That first block spoke to me. I sat in the snow as it told an entire chapter of the Inuit's amazing saga. When the wind blows across this land, the need for immediate shelter is extreme. The travelers look around, but there are no sheltering forests, only snow, and more snow, all seemingly the same for as far as the eye can see. But if they are keenly attuned to every nuance of nature, they will see a living and logical process. The snow lies in different depths with changing pressures, has varying amounts of moisture, and is packed or loosened by the prevailing winds and fluctuating temperatures. To be able to find the exact intersection of these forces is to survive in the Arctic. For here is found a building material that is plentiful, easily shaped, surprisingly strong, yet incredibly light. It allows hurried construction of a shelter that is efficiently insulated and totally windproof. This structure rises easily out of the snows when one needs it, and is just as easily left behind. In the wake of the Inuit rest no pyramids or Taj Mahals, for spring erases their efforts like footprints washed by the surf. But ask any engineer what building shape is considered the most sophisticated, and he or she will tell you it is the spiralling dome, for it requires no beams, struts, or supports. It is eloquently integrated into itself, much like its Inuit inventors. . . . With a little trial and a lot of error, I constructed a crude igloo.[205]

I would confidently award Simon an A-plus for effort. In "mimicking" the Inuk's knowledge of snow—drawn in this case from a dusty old book his cat, Halifax, slept on—he fashioned a passable igloo, albeit a wobbly one. I once tried building an igloo; it was a hopeless failure. After several cave-ins I resorted to a blue plastic tarp in lieu of a snowy ceiling.

But that doesn't mean I can't strive to be a snow master in my own little kingdom. For instance, quinzhees are my forté, not igloos. In favoring this kind of winter shelter, I make best use of the particular kind of snow that carpets my neck of the north woods. Where I live the snow is generally too soft to carve into anything you'd dare to sleep in without risking collapse at the wrong puff of wind. On the other hand, fifteen young cub scouts once posed for a picture on *top* of one of my more substantial quinzhees. So here's the thing: if you want to get into snow—I mean really fall for it—the best place to start is not necessarily on some corniced-draped mountain or the wind-flailed arctic tundra, but right where your boots hit the snow as they leave your front step.

For this sculptor, snow offers a frozen wellspring of creativity.
Fran Hurcomb

Snow fuels stories, myths, and dreams in admirers of all ages. Snow bears at play during Yellowknife's Caribou Carnival.

Fran Hurcomb

For Yellowknife's Snow King, Tony Foliot, snow provides the stock and trade of his winter livelihood—entertaining tourists and kids in his frozen kingdom.

Northern News Service

For these winter campers, snow sets the stage for a uniquely peaceful retreat into the wilderness.
Government of Northwest Territories

Not far from my door I can watch fellow citizens of my winter city connecting with snow in countless colorful ways. Take Tony Foliot, known to just about everyone in town simply as the Snow King. He has converted his deep affection for snow into an unusual business that no chamber of commerce could pigeonhole. Each winter he and his buddies build a labyrinthine snow castle beside his blue and yellow houseboat on Great Slave Lake. He then invites tourists, school groups, and locals in for tours, plays, film festivals, and art shows—all centered on the theme of snow. Including a skating rink, hair-raising ice slides, and the Little Icicles Theatre, the castle grounds cover the equivalent of half a city block.

"It's like building a sand castle," says Foliot. "I work on a snow castle for three months and then it melts back into the lake."[206]

Just over the hill from the Snow King's castle, vanloads of snow sculptors gather for Caribou Carnival every "spring" (most southerners, probably all of them actually, would still call it winter). Their work and pleasure is to give life—although a short one—to the beings they see buried in big hunks of compacted snow: winged dragons, dancing bears, leering shamans.

Then there's winter camping. On any winter weekend, even when it's −30°C (−22°F), I'll often hear my neighbors clattering around in their sheds, pulling out billy pots and sleeping bags as they prepare for a backwoods

234

camping expedition in the snow. "There's absolutely nothing like it," said a friend, as he described falling asleep on a bed of spruce boughs spread over the snow and under the stars.

And I already told you about Miles Davis and his merry band of kite-skiers. My adrenalin surges just watching them from my living room window. From there, I've even observed the odd Bombardier bug bouncing over the corrugated drifts of Back Bay—though I must admit, I've never been remotely tempted to flag one down for a joyride. In one's passion for winter pursuits, it helps to know where to draw the line.

Snow: it's worth a second look.
Tessa Macintosh

"Winter," says Thoreau, "is thrown to us like a bone to a famishing dog."[207] We can choose to slam the door on winter's leaner, meaner side and spin a protective cocoon while pining for spring. Or, as Thoreau advises, we can sink our teeth into winter and "get the marrow out of it" by befriending its richest expression: snow. The paths to greater intimacy with snow are as varied as its fresh-fallen crystals.

But remember this: no matter what path you choose, in building this relationship the sky's the limit. After all, that's where the whole story starts.

For this skier, a fresh fall of snow provides an open book to read the comings and goings of wildlife.

Tessa Macintosh

For this young snowboarder, snow offers a magic carpet ride to improved skills, ele-vated self-esteem, and supreme kicks.

Tessa Macintosh

For my daughter and my dog, snow and play are synonymous.
Tessa Macintosh

Tessa Macintosh

How to Build a Quinzhee Snowhouse

How It Works: The "Quinzhee Effect"

Unless you were born in Igloolik or Resolute in Canada's High Arctic, a quinzhee (also spelled *quinze*) is a heck of a lot easier to build than an igloo. This kind of snow shelter—and there are many other kinds, from drift burrows to ice palaces—is made by heaping snow into a large pile and hollowing out the interior. Honest, that's it.

A quinzhee holds together thanks to differences in the temperature and shape of snow crystals in the various layers of undisturbed snow on the ground. As you shovel or bucket this snow into a pile, these crystals get all stirred up, now resting much closer together. This mixing and compaction process dramatically increases their density and hardness by welding different kinds of snow together. Moister, warmer layers of snow near the ground fuse with colder crystals near the surface. Anywhere from 0.5-2 hours of settling and recrystallization are needed to cement the snow pile into one solid mass, which can then be hollowed out to make a shelter. The colder the air temperature, the greater the temperature variation in the snow. This condition in turn promotes more recrystallization and quicker cementing. I'll admit there's a bit of an art to this. If you excavate too soon, the mound may collapse on you. If you allow it to set too long, say a couple days at −35°C,

(−31°F) the snow gets so firm you'll have a heck of a time digging it out. I can speak from experience on this.

Snow fans hollow out the inside of a quinzhee snowhouse. Note the stick or "depth gauge" protruding from the ceiling.
Northern News Service

Design Features: Yours for Free
- Dome shape for structural stability
- Thick walls for maximum insulation
- Vent hole for excess moist air to escape and for good oxygen circulation

- Smooth interior walls to prevent drip points
- Raised platform to take advantage of warmer air above
- Small, low entrance hole to prevent escape of warm air
- Entrance hole positioned perpendicular to prevailing wind for protection from direct drafts and excessive drifting

Construction: Pile Up, Then Pile In

All you need to build a quinzhee is a shovel, bucket, pot, or anything else that can be used to scoop up snow and toss it into a pile. In a pinch, your own two hands will do. You will also need several sticks, each about 30 centimeters (12 inches) long.

Choose a site where the snow has not been previously disturbed. Trample or clear the snow from a circular area of about 2–4 meters (6.5–13 feet) in diameter, depending on how many people, dogs, etc. you plan to accommodate (My wife once built an eight person quinzhee; now *that* I would like to have seen.). Then pile loose snow onto the circular area to make a dome-shaped mound about 2 meters (6.5 feet) high. Make sure the snow gets thoroughly mixed by stirring it around occasionally or flinging it well into the air as you pile it. Break all blocks into fine powder and avoid any icy layers. You want a pile high and wide enough so that, when hollowed out, you can at least undress in a sitting position without touching the walls. If you want to practice your dance steps in there or hold a cello recital, well, keep piling. But keep in mind that the greater the interior volume, the longer it will take to warm up your quinzhee.

Why the sticks? They serve as depth gauges in the quinzhee's wall so that during the hollowing-out process you leave sufficient thickness for adequate insulation, at least 20 centimeters (8 inches). Push the sticks two-thirds of the way into the snow pile, spacing them evenly across the surface. Some people use lots of sticks. Others pride themselves in using only a handful. Either way, it should look like a giant pin cushion when you are finished this stage.

I usually leave the mound to harden for at least 2–3 hours. The colder it is outside and the longer you wait before digging it out, the harder your quinzhee will be. Unless you're in an emergency situation, try waiting overnight to get good strong walls.

Then, with a sturdy bowl or small shovel, begin burrowing a low entrance into the pile at ground level. Keep the entrance just big enough for one person to crawl in. It's important to start hollowing *upward* as soon as you can to reduce the potentially dangerous snowload overhead. Scrape out the lower walls and floor last. When you encounter sticks, you will know to stop burrowing in that particular direction. Thin weak walls are dangerous and just won't hold the heat. Besides, unintended windows serve no useful purpose in a quinzhee. If you really want to get fancy, you could slip in a modest pane of lake ice to let the sunshine (or moonshine) in.

To get the greatest benefit and enjoyment from your quinzhee, build a raised sleeping platform that's higher than the level of the entrance. This creates a kind of "cold well" in lower sections, allowing the heavier cold air to drain off. In your zeal for well digging, don't scrape all the snow off the quinzhee floor. Although snow can never rise above zero (who wants a slush floor?), it is still more insulative and reflective than bare ground. Keep about 15 centimeters (6 inches) of packed snow on the quinzhee floor. You might throw a caribou skin or old rug over it while you at it. Evergreen boughs or dry grass add a nice insulating touch for the pure at heart. Now is that luxury or what?

Other important finishing touches include smoothing the interior walls to eliminate potential drip points, and poking a small vent hole through the top to ensure a supply of fresh air. If you are planning to spend the night inside or are leaving the quinzhee for some time, you may want to block the entrance with a slab of snow (if conditions permit), some other flat object (I use an old piece of plywood), or a blanket.

A quinzhee remains at its best when the temperature is −15°C (5°F) or colder and the air is relatively dry. Under these conditions, a well-built quinzhee will last you an entire winter.

In *Northern Bushcraft,* Alberta outdoorsman Mors Kochankski extols the virtues of the properly built snow shelter. "The properly built snow shelter has many ideal features. The shelter is well-insulated with a low thermal mass that is easily warmed. It is as dry as any shelter can be under the circumstances and is sound and windproof. The shelter can be constructed to provide more or less room, depending on immediate needs. The interior of the shelter is bright and psychologically uplifting."[208]

Well said, Mors. Have fun out there, and enjoy your sanctuary in the snow.

Bibliography

Abbey, E. *Hayduke Lives!* New York: Little, Brown and Company, 1990.

Ahluwalia, A. "The National" CBC-TV, 14 January, 1999.

Allen, M. L. "First Snow." In *Snowy Day,* edited by C. Bauer. New York: Harper Trophy, 1986.

Anderson, L., editor. *Sisters of the Earth: Women's Prose and Poetry About Nature.* New York: Vintage, 1991.

Anonymous, "Make Your Own Snow" *Winter Living* October, 2000: 18.

Arsenault, A. "The National" CBC-TV, 14 January, 1999.

Bastedo, J. D. *Reaching North: A Celebration of the Subarctic.* Calgary: Red Deer Press, 1998.

Bastedo, J.D. "Sky Sharks, Bulldogs and Snow Fleas – Surprising Secrets of Northern Insects." *Up Here* 1993, 9(3): 62–68

Bastedo, J.D. "White Fury." *Up Here* 1999, 15(8): 62–68

Bentley, W. A., and W. J. Humpheys. *Snow Crystals.* New York: Dover Publications, 1962.

Berkes, F. *Sacred Ecology: Traditional Knowledge and Resource Management.* Philadelphia: Taylor and Francis, 1999.

Berry, W. "The Peace of Wild Things." In *Openings.* New York: Harcourt Brace Jovanovich, 1968.

Blanchard, D. *The Snowflake Man: A Biography of Wilson A. Bentley.* Blacksburg, VA: McDonald and Woodward, 1998.

Bolles, F. *Land of the Lingering Snow: Chronicles of a Stroller in New England from January to June.* Boston: Houghton Mifflin, 1891.

Boyd, D., editor. *Northern Wild: Best Contemporayr Nature Writing.* Vancouver: Greystone Books, 2001.

Boyd, D. *Rolling Thunder.* New York: Dell Publishing Co., 1974.

Butala, S. *The Perfection of the Morning: An Apprenticeship in Nature.* Toronto: HarperCollins, 1994. All excerpts used with permission of the publisher; © 1994 by Sharon Butala. All rights reserved.

Byrd, R. E. *Alone.* New York: G. P. Putnam's Sons, 1938.

Callaghan, M. *They Shall Inherit the Earth.* Toronto: McClelland and Stewart, 1962.

Colombo, J. R., ed. *Poems of the Inuit.* Ottawa: Oberon Press, 1981.

Conover, G., and A. Conover. *A Snow Walker's Companion.* Camden, ME: Mountain Press, 1995.

Cormier, M. *The Monthly Mountain.* Calgary Mountain Equipment Co-op Newsletter, March 2001.

Doré, J. "Montreal." *4th Northern Intercity Conference Report,* Tromsø, Norway: Goverment of Norway.

Dunn, John. "The Countdown from Light to Dark." In *Writing North: An Anthology of Contemporary Yukon Writers,* edited by E. Friis-Baastad and P. Robertson. Whitehorse: Beluga Books, 1992.

Dyson, J. L. *The World of Ice.* New York: Alfred Knopf, 1962.

Eiseley, L. *The Immense Journey: An Imaginative Naturalist Explores the Mysteries of Man and Nature.* New York: Random House, 1959.

Eiseley, L. "The Angry Winter." In *The Unexpected Universe.* New York: Harcourt, 1968. All excerpts used with permission of the publisher; © 1968 by Loren Eiseley and renewed 1996 by John A. Eichman, III. All rights reserved.

Emerson, R. W. "Address to the Harvard Divinity School." Cited in *Snow in America,* by B. Mergen. Washington: Smithsonian Institution Press, 1997.

Emerson, R. W. "The Snow-Storm." *In Poems of Nature,* by G. Harvey. New York: Outlet Book Company, 1989.

Ferber, P., ed. *Mountaineering: The Freedom of the Hills.* Seattle: Mountaineers, 1977.

Fitch, A. "Winter Insects of Eastern New York." *American Journal of Science and Agriculture* 5 (1847).

Flahey, B., artist. "Arthropod Iditarod International Sled Race." Artwork © 1995 by Barry Flahey, PO Box 298, Manotick, ON, K4M 1A3; www.magma.ca/~bflahey/watercolours.htm

Formozov, A. N. *Snow Cover as an Integral Factor of the Environment and Its Importance in the Ecology of Mammals and Birds.* 1946. Reprint. Edmonton: Boreal Institute, University of Alberta, 1965. (Original edition published in Materials for Fauna and Flora of the USSR. New Series Zoology 5, no. 10 (1946); translated from Russian by W. Prychodko and W. O. Pruitt).

Friis-Baastad, E., *The Exile House.* Cliffs of Moher, Ireland: Salmon Publishing, 2001. All poems by Erling Friis-Baastad used by permission of the author.

Friis-Baastad, E., and P. Robertson, eds. *Writing North: An Anthology of Contemporary Yukon Writers.* Whitehorse: Beluga Books, 1992. All poems by Erling Friis-Baastad used by permission of the author.

Frost, R. *New Enlarged Anthology of Robert Frost's Poems.* New York: Pocket Books, Simon and Schuster, 1971. ©1971 by Louis Untermeyer and Mary Silva Cosgrave.

Gadd, B. *Handbook of the Canadian Rockies.* Jasper, AB: Corax Press, 1995.

Gardner, J., editor. *The Sacred Earth; Writers on Nature and Spirit.* Navato, California: New World Library, 1998.

Goldsworthy, A. *Touching North.* London: Fabian Carlsson Gallery and Graeme Murray, Edinburgh, 1990.

Government of Northwest Territories, "Book of Chipewyans", *The Book of Dene.* Yellowknife, N.W.T.: Department of Education, 1976.

Gibson, D. "Snow Castle King." *Up Here* (March 2001): 52–54.

Gibson, W. "A Winter Walk." *Harper's* 70 (December 1885).

Gray, D. M., and D. H. Male, eds. *Handbook of Snow: Principles, Processes, Management & Use.* Willowdale, ON: Pergamon Press, 1981.

Green, A. "December Storm." Cited in *Winter: A Natural History,* by D. Sadler. Peterborough, ON: Broadview Press, 1990.

Grove, F. P. "Snow." In *Over Prairie Trails.* Toronto: McClelland and Stewart, 1957.

Haines, J. *The Stars, The Snow, The Fire: Twenty-five Years in the Alaska Wilderness.* St. Paul: Graywolf, 2000.

Harvey, G., editor. *Poems of Nature.* New York: Outlet Book Company, 1989.

Haynes, W., and J. L. Harrison, eds. *Winter Sports Verse.* New York: Duffield, 1919.

Hoham, R. "Environmental Influences on Small Algal Microbes", 58th Annual Meeting: Western Snow Conference. Sacramento, California: Western Snow Conference, 1990. 78–81.

Houston, J. *Songs of the Dream People: Chants and Images from the Indians and Eskimos of North America.* New York: Atheneum, 1972.

Hume, M., with H. Thommasen. *River of the Angry Moon: Seasons on the Bella Coola.* Vancouver: Greystone Books, Douglas and MacIntyre, 1998. All excerpts reprinted with permission of the publisher; ©1998 by Mark Hume.

Hummel, M. *Wintergreen: Reflections from Loon Lake.* Toronto: Key Porter, 1999. All excerpts used with permission of the author.

Ingram, J. *The Science of Everyday Life.* Markham, Ontario: Penguin Books, 1989.

Jones, A. *The Jerusalem Bible.* Garden City, New York: Doubleday, 1966.

Jones, H. G., J. W. Pomeroy, D. A. Walker, and R. W. Hoham, eds. *Snow Ecology: An Interdisciplinary Examination of Snow-Covered Ecosystems.* New York: Cambridge University Press, 2001.

Kearns Goodwin, D. *No Ordinary Time: Franklin and Eleanor Roosevelt: The Home Front in World War II.* New York: Touchstone Books, Simon and Schuster, 1997.

Kenyon, K. *The Boat of Quiet Hours.* St. Paul: Graywolf Press, 1986.

Kirk, R. *Snow.* Seattle: University of Washington Press, 1977.

Kochankski, M. *Northern Bushcraft.* Edmonton: Lone Pine Publishing, 1987.

Krakauer, J. *Into the Wild.* New York: Doubleday, 1996.

Kucera, R. *Exploring the Columbia Icefield.* Calgary: High Country Colour, 1999.

LaChapelle, D. *Earth Wisdom.* Los Angeles: LA Guild of Tudor Press, 1978.

LaChapelle, E. R. *A Field Guide to Snow Crystals.* Vancouver: J. J. Douglas, 1969.

LaChapelle, E. R. *Secrets of the Snow: Visual Clues to Avalanche and Ski Conditions.* Seattle: University of Washington Press, 2001.

Langford, C. "Dome Sweet Dome" *Winter Living* October 1999, 44–48 & 60.

Lastman, M. "The National" CBC-TV, 13 & 14 January, 1999.

London, J. *White Fang.* New York: Viking, 1933.

Longfellow, H. W. "Afternoon in February." In *Poems of Nature,* edited by G. Harvey. New York: Outlet Book Company, 1989.

Lopez, B. *Arctic Dreams: Imagination and Desire in a Northern Landscape.* New York: Bantam, 1986.

Lunn, P. *The Guinness Book of Skiing.* Enfield, U.K.: Guiness Superlatives, 1983.

MacDonald, J. *The Arctic Sky: Inuit Astronomy, Star Lore, and Legend.* Toronto: Royal Ontario Museum, 1998.

McKay, G. A. "Snow and Man." In *Handbook of Snow: Principles, Processes, Management & Use,* edited by D. M. Gray and D. H. Male. Willowdale, ON: Pergamon Press, 1981.

McKay, G. A., and W. P. Adams. "Snow and Living Things." In *Handbook of Snow: Principles, Processes, Management & Use,* edited by D. M. Gray and D. H. Male. Willowdale, ON: Pergamon Press, 1981.

McKelvey, B. *Snow in the Cities: A History of America's Urban Response.* Rochester, NY: University of Rochester Press, 1995.

Mergen, B. *Snow in America.* Washington: Smithsonian Institution Press, 1997. All excerpts used with permission of the author.

Moss, J. *Enduring Dreams: An Exploration of Arctic Landscape.* Concord, ON: Anansi Press, 1994.

Mowat, F. *The Snow Walker.* Toronto: McClelland and Stewart, 1975.

Muir, J. *The Yosemite.* New York: Century Co., 1912.

Murphy, C. "In Praise of Winter." *Atlantic Monthly.* January, 1995: 45-58.

Murphy, R. "The National" CBC-TV, 8 January, 1999.

Nakaya, U. *Snow Crystals: Natural and Artificial.* Cambridge, Mass.: Harvard University Press, 1954.

Norse, E. *Ancient Forests of the Pacific Northwest.* Washington: Island Press, 1990.

Olson, S. *The Singing Wilderness.* New York: Alfred A. Knopf, 1957.

Owen, N. "Breaking Trail." In *Six Windows Of The Giant.* Sault Ste. Marie, ON: Trabarni Productions, 1994. Poem used with permission of the author.

Pasternak, J. "Sweeping Snow Solutions." *Financial Post.* January 12, 2002: IN3.

Peck, M. S. *The Friendly Snowflake: A Fable of Faith, Love and Family.* Atlanta: Turner Publishing, 1992.

Pelly, D. F. *Sacred Hunt: A Portrait of the Relationship Between Seals and Inuit.* Vancouver: Greystone Books, 2001. All excerpts used with permission of the author.

Pettigrew, C. J. "Range Horses." In *Writing North: An Anthology of Contemporary Yukon Writers,* edited by E. Friis-Baastad and P. Robertson. Whitehorse: Beluga Books, 1992.

Pollock, J. S. "Winter Storms of the Century." *Winter Living,* Winter 1999, 26-31, 62-63.

Pomeroy, J. W. *Snowcover: Accumulation, Relocation and Management.* NHRI Science Report, no. 7. Saskatoon: National Hydrology Research Institute, 1995.

Poole, E. and D. Tetley. "Late storm delivers 40-cm drubbing, with more in store." *Edmonton Journal* 7 May 2002. A1 & A16.

Pruitt, W. O. *Boreal Ecology.* Institute of Biology's Studies in Biology, no. 91. London: Edward Arnold Publishers, 1978.

Pruitt, W. O. "Snow as a Factor in the Ecology of Caribou." *Arctic* 12(3) 1959: 171-72. All excerpts used with permission of the author.

Pruitt, W. O. *Wild Harmony: The Cycle of Life in the Northern Forest.* Saskatoon: Western Producer Prairie Books, 1983. All excerpts used with permission of the author.

Pryde, D. Nunaga: *Ten Years of Eskimo Life*. New York: Walker and Company, 1971.

Raymo, C. *Honey From Stone*. Kerry, Ireland: Dodd, Dingle, Co., 1997.

Rogers, W. C. "Chips, Flakes and Gusts." *Livable Winter Newsletter* 5(2) June 1987: 5-6.

Sadler, D. *Winter: A Natural History*. Peterborough, ON: Broadview Press, 1990.

Sandford, R. W. *The Columbia Icefield*. Banff, AB: Altitude Publishing, 1993.

Schmidt, G. D., ed. *Poetry for Young People: Robert Frost*. New York: Scholastic, 1994.

Schulman, J. F. ed. *Ralph Waldo Emerson Speaks*. Boston: Unitarian Universalist Association, 2000.

Scoresby, W. *An Account of the Arctic Regions*. Vol. 1. Edinburgh: Edinburgh Press, 1820.

Seligman, G. *Snow Structure and Ski Fields*. London: Macmillan and Co., 1936.

Service, R. *Collected Poems of Robert Service*. Toronto: McGraw-Hill Ryerson, 1960.

Simon, A. *North to the Night: A Spiritual Odyssey in the Arctic*. New York: Broadway Books, 1998.

Sims, T. *The Treasures of the Snow*. Toronto: William Briggs, 1896.

Stanwell-Fletcher, T. *Driftwood Valley*. New York: Little, Brown and Co., 1946. Copyright renewed 1974 by Theodora C. Stanwell-Fletcher. Used by permission of Little, Brown and Company (Inc).

Stein, S. *Stein on Writing*. New York: St Martin's Griffin, 1995.

Steinwand, V. "Bugs on ice." *News North* 9 April 2001. 1.

Stevens, W. "The Snow Man." 1921. Reprint in *News of the Universe: Poems of Twofold Consciousness,* edited by R. Bly. San Francisco: Sierra Club Books, 1980.

Struzik, E. "The weather doctor is in." *Edmonton Journal* 9 Dec. 2001. D12.

Swenson, M. "Snow by Morning." Cited in *Snow in America,* by B. Mergen. Washington: Smithsonian Institution Press, 1997.

Thoreau, H. D. Ktaadn. Cited in *Into the Wild,* by J. Krakauer. New York: Doubleday, 1996.

Tutton, A. E. *The High Alps: A Natural History of Ice and Snow*. London: Kegan Paul, Trench, Trubner, 1927.

Tyndall, J. *The Forms of Water in Clouds & Rivers, Ice & Glaciers*. London: Henry S. King, 1872.

Washington, J. *Eskimaux and English Vocabulary for the Use of the Arctic Expeditions*. London: J. Murray, 1850.

Watson, J. "Beautiful Snow." Cited in *The Treasures of the Snow,* by T. Sims. Toronto: William Briggs, 1896.

Williams, T. T., and Major, T. *The Secret Language of Snow*. San Francisco: Sierra Club/Pantheon Books, 1984.

Yates, M. "The Hunter Who Loses His Human Scent." *Canadian Fiction Magazine* 83: Arctic of Words (1993): 127-36.

Endnotes

1. Duncan Blanchard, *The Snowflake Man: A Biography of Wilson A. Bentley* (Blacksburg, Virginia: McDonald & Woodward, 1998) 88.
2. Cited in: Loren Eiseley, *The Unexpected Universe* (New York: Harcourt, 1968) 97.
3. Cited in: Jon Krakauer, *Into the Wild* (New York: Doubleday, 1996) 172.
4. Cited in: Richard Kucera, *Exploring the Columbia Icefield* (Calgary: High Country Colour, 1999) 7.
5. Kucera, 22.
6. Kucera, 11.
7. Cited in: Jason Gardner, editor, *The Sacred Earth - Writers on Nature and Spirit* (Novato, California: New World Library, 1998) 51.
8. Bernard Mergen, *Snow in America* (Washington: Smithsonian Institution Press, 1997) 239.
9. Alexander Jones, editor, *The Jerusalem Bible* (Garden City, New York: Doubleday, 1966) 671.
10. Thomas Sims, *The Treasures of the Snow* (Toronto: William Briggs, 1896) 13.
11. Gray and Male, editors. *Handbook of Snow - Principles, Processes, Management & Use* (Willowdale, Ontario: Pergamon Press, 1981) 129.
12. A.E. Tutton, *The High Alps - A Natural History of Snow and Ice* (London: Kegan Paul, Trench, Trubner, 1927) 1-2.
13. J. Tyndall, *The Forms of Water in Clouds & Rivers, Ice & Glaciers* (London: Henry S. King, 1872) 31-32.
14. Tyndall, 34.
15. Duncan Blanchard, *The Snowflake Man: A Biography of Wilson A. Bentley* (Blacksburg, Virginia: McDonald & Woodward, 1998) 172-173.
16. Ulkichiro Nayaka, *Snow Crystals: Natural and Artificial* (Cambridge, Mass.: Harvard University Press, 1954) 4.
17. Jay Ingram, *The Science of Everyday Life* (Markham, Ontario: Penguin Books) 153.
18. Mergen, 197.
19. Mergen, 197.
20. Mergen, 197
21. Chet Raymo, *Honey From Stone* (Dingle, Co. Kerry, Ireland: Brandon, 1997) 45.

22. Raymo, 46.

23. Raymo, 47.

24. Gray and Male, 143.

25. Cited in: G. Seligman, *Snow Structure and Ski Fields* (London: MacMillan & Co., 1936) 36.

26. Blanchard, 60.

27. Blanchard, 60-61.

28. Seligman, 41.

29. Ingram, 156.

30. Cited in: Blanchard, 174.

31. Mergen, xxi.

32. Arthur Black, CBC radio interview, 15 March 1998.

33. Seligman, 44.

34. A.E. Tutton, *The High Alps: A Natural History of Ice and Snow* (London: Kegan Paul, Trench, Trubner, 1927) 4.

35. Mergen, 249.

36. Peggy Ferber, editor. *Mountaineering: The Freedom of the Hills* (Seattle, Washington: The Mountaineers, 1977) 245 & 390.

37. Raymo, 52.

38. Blanchard, 51.

39. Mergen, 205.

40. Edward Abbey, *Hayduke Lives!* (New York: Little, Brown & Company, 1990) 128.

41. Abbey, 128.

42. Abbey, 128.

43. Cited in: Jon Krakauer, *Into the Wild* (New York: Doubleday, 1996) 9.

44. William Pruitt, personal correspondence, 10 Feb. 1997.

45. W. O. Pruitt, *Wild Harmony: The Cycle of Life in the Northern Forest* (Saskatoon: Western Producer Prairie Books, 1983) 34.

46. Barry Lopez, *Arctic Dreams - Imagination and Desire in a Northern Landscape* (New York: Bantam, 1986) 197 & 198.

47. Cited in: Mergen, 32.

48. Jamie Bastedo, "Sky Sharks, Bulldogs and Snow Fleas - Surprising Secrets of Northern Insects." *Up Here* (April 1993) 62.

49. H. Gerald Jones, John W. Pomeroy, D.A. Walker and Ronald.W. Hoham, editors. *Snow Ecology: An Interdisciplinary Examination of Snow-Covered Ecosystems* (New York: Cambridge University Press, 2001) 238.

50. Ben Gadd, *Handbook of the Canadian Rockies* (Jasper, Alberta: Corax Press, 1995) 533.

51. Cited in: Mergen, 184.

52. Cited in: Mergen, 32.

53. Robert Service, *Collected Poems of Robert Service* (Toronto: McGraw-Hill Ryerson Limted, 1960) 630-635.

54. Gadd, 533.

55. James Dyson, *The World of Ice* (New York: Alfred Knopf, 1962) 138.

56. Ronald Hoham, "Environmental Influences on Small Algal Microbes", *58th Annual*

Meeting: Western Snow Conference (Sacramento, California: Western Snow Conference, 1990) 78.

57. Mergen, 111.

58. Hoham, 81.

59. Cited in: Jamie Bastedo, *Reaching North: A Celebration of the Subarctic* (Calgary: Red Deer Press, 1998) 87.

60. Cited in: Bastedo, 88.

61. Cited in: Bastedo, 95–96.

62. William O. Pruitt, "Snow as a Factor in the Ecology of Caribou", *Arctic* (1959 12[3]) 175.

63. Pruitt, "Snow as a Factor in the Ecology of Caribou", 173.

64. Pruitt, "Snow as a Factor in the Ecology of Caribou", 171–172.

65. William Pruitt, personal correspondence, 28 July 2002.

66. Gray and Male, editors. *Handbook of Snow - Principles, Processes, Management & Use* (Willowdale, Ontario: Pergamon Press, 1981) 14 & 16.

67. Cited in: Bernard Mergen, *Snow in America* (Washington: Smithsonian Institution Press, 1997) 31–32.

68. Bernard Mergen, *Snow in America* (Washington: Smithsonian Institution Press, 1997) 249.

69. Cited in: Caroline Bauer, editor, *Snowy Day* (New York: Harper Trophy, 1986) 24.

70. M. Scott Peck, *The Friendly Snowflake - A Fable of Faith, Love and Family* (Atlanta: Turner Publishing, 1992) 30.

71. Sigurd Olson, *The Singing Wilderness* (New York: Alfred A. Knopf, 1957) 192.

72. John Dunn, "The Countdown from Light to Dark" in Erling Friis-Baastad & Patricia Robertson, editors, *Writing North - An Anthology of Contemporary Yukon Writers* (Whitehorse: Beluga Books, 1992) 3.

73. Cited in: Gary D. Schmidt, *Poetry for Young People: Robert Frost* (New York: Scholastic, 1994) 33.

74. C. J. Pettigrew, "Range Horses" in Friis-Baastad & Robertson, 169.

75. Cited in: Mergen, 228.

76. Cited in: Mergen, 215.

77. Cited in: Doug Sadler, *Winter: A Natural History* (Peterborough, Ontario: Broadview Press, 1990) 51.

78. Robert Service, *Collected Poems of Robert Service* (Toronto: McGraw-Hill Ryerson Limted, 1960) 178.

79. Cited in: Thomas Sims, *The Treasures of the Snow* (Toronto: William Briggs, 1896) 13–15.

80. Cited in: George Woodcock, "Terror and Regeneration: The Wilderness in Art and Literature", in Borden Spears, editor. *Wilderness Canada* (Toronto: Clarke, Irwin, 1970) 89.

81. William Gibson, "A Winter Walk." *Harper's* (December 1885) 70.

82. Cited in: Schmidt, 37.

83. Frederick Philip Grove, *Over Prairie Trails* (Toronto: McClelland & Stewart, 1957).

84. Loren Eiseley, *The Unexpected Universe* (New York: Harcourt, 1968) 118–119.

85. Monte Hummel, *Wintergreen - Reflections from Loon Lake* (Toronto: Key Porter, 1999) 14-16.
86. Gail Harvey, editor, *Poems of Nature* (New York: Outlet Book Company, 1989) 12.
87. Sharon Butala, *The Perfection of the Morning - An Apprenticeship in Nature* (Toronto: HarperCollins, 1994) 84-85.
88. Cited in: Mergen, 229.
89. Cited in: Sol Stein, *Stein on Writing* New York: St Martin's Griffin, 1995) 228.
90. Theodora Stanwell-Fletcher, *Driftwood Valley* New York: Little Brown & Company, 1946. Cited in: Lorraine Anderson, *Sisters of the Earth* (New York:Vintage, 1991) 61.
91. James Houston, *Songs of the Dream People - Chants and Images from the Indians and Eskimos of North America* (New York: Atheneum, 1972) 63.
92. Service, 12.
93. Mark Hume with Harvey Thommasen, 1998. *River of the Angry Moon: Seasons on the Bella Coola.* Vancouver: Greystone Books. Cited in: David Boyd, editor, *Northern Wild: Best Contemporary Canadian Nature Writing* (Vancouver, Greystone Books, 2001) 263 & 271.
94. Robert Frost, *New Enlarged Anthology of Robert Frost's Poems* (New York: Pocket Books, 1971) 194.
95. Andy Goldsworthy, *Touching North* (London: Fabian Carlsson Gallery and Graeme Murray, Edinburgh, 1990).
96. David Pelly, *Sacred Hunt* (Vancouver: Greystone Books, 2001) 45-46.
97. John Muir, *The Yosemite* (New York: The Century Co., 1912) Cited in: Lee Stetson, editor, *The Wild Muir: Twenty-two of John Muir's Greatest Adventures* (Yosemite National Park, California:Yosemite Association, 1994) 55 & 61-65.
98. Erling Friis-Baastad, "Spending you death in the Yukon", *The Exile House* (Cliffs of Moher, Ireland: Salmon Publishing, 2001) 28.
99. Frost, 183.
100. Cited in: Schmidt, 35.
101. Houston, 75.
102. Sims, 16.
103. John R. Colombo, editor, *Poems of the Inuit* (Ottawa: Oberon Press, 1981) 47.
104. Cited in: Robert Bly, editor, *News of the Universe - Poems of Twofold Consciousness* (San Francisco: Sierra Club Books, 1980) 115.
105. Cited in: Jon Krakauer, *Into the Wild* (New York: Doubleday, 1996) 127.
106. Dolores LaChapelle, *Earth Wisdom* (Los Angeles: L.A. Guild of Tudor Press, 1978).
107. Doug Boyd, *Rolling Thunder* (New York: Dell Publishing, 1974). 72.
108. Sims, 7.
109. J. Frank Schulman, editor, *Ralph Waldo Emerson Speaks* (Boston: Unitarian Universalist Association, 2000) 1.
110. Cited in: Mergen, 10.
111. John Moss, *Enduring Dreams - An Exploration of Arctic Landscape.* Concord, Ontario: Anansi, 1994) 77.
112. Michael Yates, "The Hunter Who Loses His Human Scent", *Canadian Fiction Magazine* (Number 83: Arctic of Words, 1993) 127-136.

113. David Pelly, *Sacred Hunt* (Vancouver: Greystone Books, 2001) 60.

114. Cited in: Pelly, 71.

115. Frost, 176.

116. Jane Kenyon, *The Boat of Quiet Hours* (Saint Paul, MN: Graywolf Press, 1986) Cited in: Anderson, 200.

117. Cited in: Elizabeth Roberts & Elias Amidon, editors, *Earth Prayers.* San Francisco: Harper Collins, 1991) 296.

118. Erling Friis-Baastad and Patricia Robertson, editors, *Writing North - An Anthology of Contemporary Yukon Writers* (Whitehorse: Beluga Books, 1992) 4-5.

119. Schmidt, 34.

120. Frost, 184.

121. Cited in: Bernard Mergen, *Snow in America* (Washington: Smithsonian Institution Press, 1997) 246.

122. Blake McKelvey, *Snow in the Cities: A History of America's Urban Response* (Rochester, N.Y.: University of Rochester Press, 1995) xiii.

123. McKelvey, 26.

124. McKelvey, 6.

125. McKelvey, 8.

126. McKelvey, 10.

127. McKelvey, 10.

128. McKelvey, 15.

129. McKelvey, 39.

130. McKelvey, 59.

131. Mergen, 36.

132. Mergen, 37.

133. Duncan Blanchard,. *The Snowflake Man: A Biography of Wilson A. Bentley* (Blacksburg, Virginia: McDonald & Woodward, 1998) 46 & 67.

134. Mergen, 37.

135. Mergen, 50.

136. Mergen, 63.

137. Mergen, 50.

138. Julie Suzanne Pollock, "Winter Storms of the Century", *Winter Living* (October, 1999) 30.

139. Pollock, 31.

140. Adrienne Arsenault, "The National" CBC-TV, 14 Jan. 1999.

141. Raj Ahluwalia, "The National" CBC-TV, 14 Jan. 1999.

142. Arsenault, 14 Jan. 1999.

143. Mel Lastman, "The National" CBC-TV, 13 Jan. 1999.

144. Mel Lastman, "The National" CBC-TV, 14 Jan. 1999.

145. Rex Murphy, "The National" CBC-TV, 8 Jan. 1999.

146. McKelvey, xiii.

147. G.A. McKay, "Snow and Man" in Gray and Male, editors. *Handbook of Snow: Principles, Processes, Management & Use* (Willowdale, Ontario: Pergamon Press, 1981).

148. Jamie Bastedo, "White Fury", *Up Here* (Nov. /Dec. 1999) 18.

149. Bastedo, 18.

150. Duncan Pryde, *Nunaga: Ten Years of Eskimo Life* (New York: Walker and Company, 1971) Cited in: Doug Sadler, *Winter: A Natural History* (Peterborough, Ontario: Broadview Press, 1990) 68.

151. Bastedo, 18.

152. Cited in: Jerry Dennis, *It's Raining Fishes and Frogs* (New York: Harper Collins, 1992) 296.

153. Bastedo, 18.

154. Mergen, 75.

155. Anonymous, "Make Your Own Snow", *Winter Living* (October 2000) 18.

156. www.backyardblizzard.com

157. McKelvey, 38.

158. Cooper Langford, "Dome Sweet Dome", *Winter Living* (October 1999) 46.

159. Langford, 46.

160. Langford, 60.

161. Jean Doré, "Montreal", *4th Northern Intercity Conference Report* (Tromsø, Norway: Government of Norway) 21.

162. James Pasternak, "Sweeping Snow Solutions", *Financial Post* (12 Jan. 2002) IN3.

163. Pasternak, IN3.

164. Pasternak, IN3.

165. Thomas Sims, *The Treasures of the Snow* (Toronto: William Briggs, 1896) 5.

166. McKelvey, viii.

167. McKelvey, xix.

168. Mergen, 74.

169. Mergen, 75.

170. Cullen Murphy, "In Praise of Snow" *Atlantic Monthly* (Jan. 1995) 47.

171. Mike Cormier, *The Monthly Mountain* (Calgary Mountain Equipment Co-op newsletter, March 2001) 1.

172. Emma Poole and Deborah Tetley, "Late storm delivers 40–cm drubbing, with more in store", *Edmonton Journal* (7 May 2002) A16.

173. Poole and Tetley, A16.

174. William C. Rogers, "Chips, Flakes and Gusts", *Livable Winter Newsletter* (June 1987) 5.

175. Rogers, 5.

176. William C. Rogers, "Chips, Flakes and Gusts", *Winter Cities News* (Feb. 1989, 7[1]) 10.

177. Ed Struzik, "The weather doctor is in", *Edmonton Journal* (9 Dec. 2001) D12.

178. Story adapted from: Government of Northwest Territories "Book of Chipewyans", *The Book of Dene* (Yellowknife, N.W.T.: Department of Education, 1976) 1.

179. Garret and Alexandra Conover, *A Snow Walker's Companion* (Camden, Maine: Ragged Mountain Press, 1995) 5.

180. Jamie Bastedo, *Reaching North: A Celebration of the Subarctic* (Calgary: Red Deer Press, 1998) 95.

181. Monte Hummel, *Wintergreen - Reflections from Loon Lake* (Toronto: Key Porter Books, 1999) 11–13.

182. Cited in: Conover, 5.

183. Neill Owen, *Six Windows Of The Giant* (Sault Sainte Marie, Ontario: Trabarni Productions, 1994).

184. Bernard Mergen, *Snow in America* (Washington: Smithsonian Institution Press, 1997) 120.

185. Cited in: Mergen, 96.

186. Cited in: Mergen, 96.

187. Cited in: Gray and Male, editors. *Handbook of Snow - Principles, Processes, Management & Use* (Willowdale, Ontario: Pergamon Press, 1981) 732.

188. Cited in: Gray and Male, 722.

189. Cited in: Gray and Male, 728.

190. Mergen, 114.

191. Mergen, 110.

192. Edward R. LaChapelle, *Secrets of the Snow: Visual Clues to Avalanche and Ski Conditions* (Seattle: Universtiy of Washington Press, 2001) 94-95.

193. Peter Lunn, *The Guiness Book of Skiing* (Enfield, U.K.: Guinness Superlatives, 1983) 99.

194. Ruth Kirk, *Snow* (Seattle: University of Washington Press, 1977) 11.

195. Verne Steinwand, "Bugs on ice", *News North* (9 April 2001) 1.

196. John MacDonald, *The Arctic Sky: Inuit Astronomy, Star Lore, and Legend* (Toronto: Royal Ontario Museum, 1998) 175.

197. MacDonald, 175.

198. Andy Goldsworthy, *Touching North* (London: Fabian Carlsson Gallery and Graeme Murray, Edinburgh, 1990). Glossary.

199. Bernard Mergen, *Snow in America* (Washington: Smithsonian Institution Press, 1997) 160.

200. Mergen, 161.

201. Fikret Berkes, *Sacred Ecology: Traditional Knowledge and Resource Management* (Philadelphia: Taylor & Francis, 1999) 46.

202. Berkes, 45.

203. MacDonald, 177.

204. John Moss, *Enduring Dreams: An Exploration of Arctic Landscape* (Concord, Ontario: Anansi Press Limited, 1994) 77.

205. Alvah Simon, *North to the Night - A Spiritual Odyssey in the Arctic* (New York: Broadway Books, 1998) 153.

206. Dane Gibson, "Snow Castle King", *Up Here* (March 2001) 52.

207. Cited in: Loren Eiseley, *The Unexpected Universe* (New York: Harcourt, 1968) 97.

208. Mors Kochankski, *Northern Bushcraft* (Edmonton: Lone Pine Publishing, 1987) 184.

J amie Bastedo's work is all about taking science to the streets. Whether playing zany environmental songs around the campfire, hosting lively nature shows on the radio, performing as an arctic explorer, leading ecotours or writing about nature, Jamie Bastedo spreads a catching enthusiasm and love for the land. Well established as a popular science writer, he has written five books on northern nature (including *Reaching North* and *Shield Country)* and many natural history features and reviews in magazines such as *Up Here, Backpacker, Winter Living,* and *Canadian Geographic.* Jamie Bastedo was awarded the Michael Smith Award for his outstanding contribution to the promotion of science as well as the Queen's Jubilee Medal for his work in environmental education and conservation. When not out on the land, he hangs his hat in Yellowknife, Northwest Territories, where he lives with his wife and daughters.